Field Speciation Method for Arsenic Inorganic Species

The mission of the Awwa Research Foundation (AwwaRF) is to advance the science of water to improve the quality of life. Funded primarily through annual subscription payments from over 1,000 utilities, consulting firms, and manufacturers in North America and abroad, AwwaRF sponsors research on all aspects of drinking water, including supply and resources, treatment, monitoring and analysis, distribution, management, and health effects.

From its headquarters in Denver, Colorado, the AwwaRF staff directs and supports the efforts of over 800 volunteers, who are the heart of the research program. These volunteers, serving on various boards and committees, use their expertise to select and monitor research studies to benefit the entire drinking water community.

Research findings are disseminated through a number of technology transfer activities, including research reports, conferences, videotape summaries, and periodicals.

Field Speciation Method for Arsenic Inorganic Species

Prepared by:
Dennis A. Clifford, **Saqib Karori**, **Ganesh Ghurye**, and **Gautam Samanta**
The University of Houston
Department of Civil and Environmental Engineering
Houston, TX 77204-4791

Jointly sponsored by:
Awwa Research Foundation
6666 West Quincy Avenue, Denver, CO 80235-3098

U.S. Environmental Protection Agency
Washington, D.C.

and

Association of California Water Agencies
Sacramento, CA

Published by:

Awwa Research Foundation
Advancing the Science of Water®

IWA Publishing

DISCLAIMER

This study was funded by the Awwa Research Foundation (AwwaRF). AwwaRF assumes no responsibility for the content of the research study reported in this publication or for the opinions or statements of fact expressed in the report. The mention of trade names for commercial products does not represent or imply the approval or endorsement of AwwaRF. This report is presented solely for informational purposes.

Published by IWA Publishing, Alliance House, 12 Caxton Street, London SW1H 0QS, UK

Telephone: +44 (0) 20 7654 5500; Fax: +44 (0) 20 7654 5555; Email: publications@iwap.co.uk;
Web: **www.iwapublishing.com**

AwwaRF report number 91014F. Originally published by AwwaRF for its subscribers in 2004.
IWA Publishing version published 2005.

ISBN 1 84339 907 5

CONTENTS

TABLES

FIGURES

FOREWORD

The Awwa Research Foundation is a nonprofit corporation that is dedicated to the implementation of a research effort to help utilities respond to regulatory requirements and traditional high-priority concerns of the industry. The research agenda is developed through a process of consultation with subscribers and drinking water professionals. Under the umbrella of a Strategic Research Plan, the Research Advisory Council prioritizes the suggested projects based upon current and future needs, applicability, and past work; the recommendations are forwarded to the Board of Trustees for final selection. The foundation also sponsors research projects through an unsolicited proposal process; the Collaborative Research, Research Applications, and Tailored Collaboration programs; and various joint research efforts with organizations such as the U.S. Environmental Protection Agency, the U.S. Bureau of Reclamation, and the Association of California Water Agencies.

This publication is a result of one of these sponsored studies, and it is hoped that its findings will be applied in communities throughout the world. The following report serves not only as a means of communicating the results of the water industry's centralized research program but also as a tool to enlist the further support of the nonmember utilities and individuals.

Projects are managed closely from their inception to the final report by the foundation's staff and large cadre of volunteers who willingly contribute their time and expertise. The foundation serves a planning and management function and awards contracts to other institutions such as water utilities, universities, and engineering firms. The funding for this research effort comes primarily from the Subscription Program, through which water utilities subscribe to the research program and make an annual payment proportionate to the volume of water they deliver and consultants and manufacturers subscribe based on their annual billings. The program offers a cost-effective and fair method for funding research in the public interest.

A broad spectrum of water supply issues is addressed by the foundation's research agenda: resources, treatment and operations, distribution and storage, water quality and analysis, toxicology, economics, and management. The ultimate purpose of the coordinated effort is to assist water suppliers to provide the highest possible quality of water economically and reliably. The true benefits are realized when the results are implemented at the utility level. The foundation's trustees are pleased to offer this publication as a contribution toward that end.

Walter J. Bishop
Chair, Board of Trustees
Awwa Research Foundation

James F. Manwaring, P.E.
Executive Director
Awwa Research Foundation

ACKNOWLEDGMENTS

The authors are indebted to the following organizations and personnel for the excellent cooperation and in-kind matching funding provided for this study:

- City of Albuquerque, New Mexico, Public Works Department: John M. Stomp III, manager, Water Resources Division; and Barbara Gastian, water operations supervisor
- City of Grand Forks, North Dakota, Department of Drinking Water: Craig Lacher, lab supervisor
- McFarland Mutual Water Company, McFarland, California: Jim Logsdon, public works director, and Kelly Ulrich, general manager (former)
- University of Minnesota, Department of Civil Engineering: Melinda (Mindy) Erickson, doctoral student, and her advisor Dr. Randal Barnes

The authors value the technical oversight and many helpful suggestions given to us by the Awwa Research Foundation (AwwaRF) Project Advisory Committee members for this project: Prof. Janet G. Hering, California Institute of Technology; Dr. John R. Garbarino, U.S. Geological Survey; Dr. John T. Creed, U.S. Environmental Protection Agency; and David E. Kimbrough, Castaic Lake Water Authority.

The authors are very grateful for the funding provided by AwwaRF and for the technical and administrative guidance provided by Dr. Djanette Khiari, the project manger.

Additional matching funding was provided by the University of Houston through the Office of Research led by Dr. Art Vailas, vice president for research. Capable administration of the grant was provided by Pat Dehoop, associate director, Office of Contracts and Grants.

EXECUTIVE SUMMARY

Inorganic arsenic (As) in drinking water is considered a human carcinogen with multiple sites of attack. There are numerous reports in the literature, based on past and ongoing experience in various countries in Asia and South America concerning the higher risks of skin, bladder, lung, liver, and kidney cancer that result from continued consumption of elevated levels of arsenic in drinking water. In ground waters, only inorganic arsenic in the form of arsenite [As III] and arsenate [As V] are significant. As III is much more toxic than As V. Moreover, due to the anionic character of As V in the drinking water, As V can be removed more easily than As III. Arsenic speciation of water samples changes over time, specifically, during transportation of samples to the laboratory and storage of samples for analysis due to redox reactions and photochemical oxidation. But, it is important to know the exact concentrations of the arsenic species in the drinking water at the source in order to develop proper treatment strategies, and for epidemiological studies. Therefore, arsenic-containing waters must be speciated immediately in the field or preserved for later speciation in the laboratory. Currently, there is no generally accepted (standard) method for field speciation or for preservation. This study describes the development of a reliable field-speciation method for inorganic arsenic species, and compares the new method with a recently developed USEPA preservation method.

RESEARCH OBJECTIVES

The overall objective of the arsenic speciation study was to develop a robust field arsenic speciation method to separate As III from As V for subsequent analysis in an analytical laboratory. To achieve this objective, the following tasks were undertaken.

1. Identify existing speciation methods that show potential for adaptation to field use.
2. Compare the relative advantages and drawbacks of the identified speciation methods.
3. Evaluate the effect of ferrous iron on the identified speciation methods.
4. Select the best method and evaluate it for accuracy, precision, and robustness.
5. Determine the fate of organic arsenic (MMA and DMA) in the speciation method.
6. Develop a field protocol for the selected method, and verify the performance of the method in the field.

PROJECT APPROACH

The project included an information-gathering phase for the existing field speciation methods. Initially, the relative merits and drawbacks of the two existing speciation media: IX-filter disks and anion-exchange resin minicolumns with both chloride- and acetate-form resins were evaluated in the laboratory using the National Sanitation Foundation Standard 53 challenge water (the NSF challenge water) without iron. To study the effect of Fe II in the speciation study, DO-free challenge water was produced and spiked with ferrous iron. The comparative performance of the accepted speciation methods (Clifford method at pH 8.3 and 3.0, and Edwards method) based on the preliminary experiments was re-evaluated in the presence of Fe II in the solution. In the presence of Fe II, the above-mentioned three methods were not acceptable for field speciation of inorganic arsenic species.

A new field speciation method using ethylenediaminetetraacetate (EDTA) and acetic acid (HAc) was developed in the laboratory, and its field effectiveness was evaluated at actual field sites where ground water samples were collected and speciated at several well sites in Ohio, California, North Dakota, and Minnesota. The concentrations of EDTA and HAc were chosen to be the same as their concentrations used in EPA's recently developed arsenic preservation method. This was done to allow both preservation and speciation of the same sample at the sampling site.

Finally, a detailed field speciation protocol for inorganic arsenic speciation was developed for a robust and accurate field speciation.

CONCLUSIONS

Due to low ion-exchange capacity, the IX filter disks both in the chloride and acetate forms were not suitable for arsenic speciation in the pH 2-3 solutions, which had significantly increased ionic strength compared with non-acidified samples. Low pH was necessary to eliminate the interference of ferrous iron.

Minicolumns with either chloride- or acetate-form resin were equally efficient for arsenic speciation in the absence of Fe II in non-acidified samples. But, arsenic speciation by minicolumns was seriously affected in acidified samples. The chloride-form resin minicolumn provided acceptable speciation performance at pH 3.0 when acidified by HCl only. It failed at pH 2.0. The acetate-form minicolumn also provided acceptable speciation performance at pH 3.0 (acidified by either HCl or H_2SO_4) but not at pH 2.0 when acidified with HCl.

In the presence of Fe II, the three existing methods—Clifford methods at pH 8.3 and pH 3.0, and the Edwards method at pH 2.0—were evaluated, and the results showed that Fe II interference was significant because it was oxidized to Fe III, which hydrolyzed and precipitated as ferric oxyhydroxides. These ferric oxyhydroxides adsorbed arsenic species and prevented their proper separation by the resin minicolumns.

The newly developed EDTA-HAc method could efficiently separate the inorganic species of arsenic in the presence or absence of Fe II. Immediate lowering of pH using HAc and complexing of iron with EDTA could remove the interference of iron efficiently. The resin volume and the speciation flow rate were optimized. At least 10 mL of chloride-form SBA resin must be used to obtain 60 mL of speciated sample. Proportionately less resin is required when less speciated sample is needed. The presence of sulfate ion up to 240 mg/L was studied and found not to interfere with arsenic speciation. The new EDTA-HAc speciation method was tested on numerous well waters by five different operators at field locations in Ohio, North Dakota, California, New Mexico, and Minnesota. When the total and speciated arsenic samples were subsequently analyzed in the lab at UH, arsenic concentration was measured by hydride generation atomic absorption spectroscopy (HG-AAS). The EDTA-HAc method was compared with the modified Battelle-Edwards method (samples analyzed by ICP-MS at Battelle Columbus) for the Ohio samples, and the comparison study showed similar speciation results. For Minnesota well waters, the EDTA-HAc preserved samples were speciated in the field by the minicolumn and also speciated in the laboratory in the presence of citric/citrate buffer at pH 5.0 by HG-AAS. There was no statistically significant difference at 95% confidence level between these two results, which indicated that the new EDTA-HAc speciation method could be used as a robust and an accurate field speciation method for inorganic species.

CHAPTER 1
INTRODUCTION

ARSENIC CHEMISTRY IN GROUNDWATERS

Arsenic contamination is primarily a ground water problem although some surface waters are also affected (Reid 1994). In ground waters, inorganic arsenite (As III) and arsenate (As V) predominate, whereas surface waters may contain significant organic arsenic including monomethylarsonic acid (MMA) and dimethylarsinic acid (DMA). The expected forms of inorganic and organic arsenic in drinking water are presented in Table 1.1. The order of expected occurrence of arsenic species in drinking water is As V > As III >> MMA and DMA. At high redox potentials, which are characteristic of oxygenated surface and some ground waters, As V was observed to be the predominant form of inorganic arsenic (Ferguson and Gavis 1972). Because the pK_2 of arsenic acid (H_3AsO_4) is 6.98, the dominant As V species in the pH range of 6-7 is monovalent anion $H_2AsO_4^-$, whereas in the pH range of 7-9, the divalent anion $HAsO_4^{-2}$ is dominant. Arsenate anions—both monovalent and divalent—can be easily concentrated from water using anion-exchange (Horng 1983; Clifford 1999). Under mildly reducing conditions, the predominant form of inorganic arsenic is arsenious acid (H_3AsO_3).

Table 1.2 summarizes some of the published literature on the distribution of the arsenic species found in ground and surface waters. In these selected studies, it has been found that inorganic forms of arsenic viz, As III and As V are dominant. For highly contaminated waters, Irgolic (1981) developed and tested analytical methods that would allow the speciation of arsenic in drinking waters from around the world. He reported no indication of the presence of organic arsenic compounds; only arsenite and arsenate were detected, and these two species accounted for the total arsenic present when the detection limit was in the low ppb range. The arsenite/arsenate ratios in the waters analyzed by Irgolic were highly variable and in a broad range of 0.007 to 3.4.

Clifford and Lin (1991) also found that arsenite and arsenate accounted for all of the arsenic present in the well waters they analyzed in Hanford, CA and San Ysidro, NM. To speciate inorganic arsenic, they used the method developed by Clifford et al. (1983). The water in a 550-ft deep Hanford well and in 44-ft deep San Ysidro test well contained a mixture of 35% As III and 65% As V, an undetectable level of iron.

In a later study in Hanford, CA, Hering and Chiu (2000) found that most of the wells surveyed contained As III (typically >89%). However, they also found that water storage sometimes facilitated the oxidation of As III to As V based on finding only 4-14% As III in one of the storage tanks.

Inorganic arsenic doesn't necessarily account for all the arsenic present in natural waters. Working with surface waters that generally contained much lower levels of arsenic than the waters analyzed by Irgolic (1982) and Clifford and Lin (1991), Andreae (1977) found small percentages of the organic arsenicals MMA and DMA. Generally, however, inorganic arsenic accounted for 85-99% of the total arsenic. Irgolic (1982) concluded that, because arsenate was the thermodynamically favored species and the presence of MMA and DMA was correlated with primary productivity, biological activity was responsible for the non-equilibrium concentrations of the organic arsenic compounds.

The organic forms of arsenic, viz. MMA and DMA, are found in waters where microbial activity is significant or where herbicide pollution by MMA or DMA has occurred

Table 1.1
Inorganic and organic arsenic compounds expected in water supplies

Compound name	Structural formula	Predominant species at pH 6.5 in natural water	Predominant species at pH 8.5 in natural water
Arsenious Acid $pK_1 = 9.22$ Inorganic As III	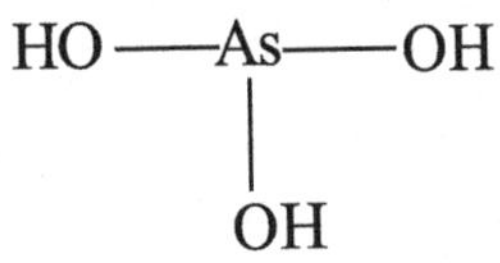	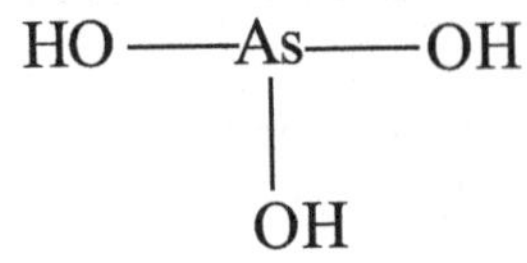 Uncharged molecule	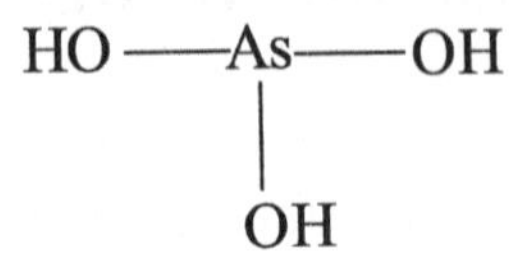 Uncharged Molecule
Arsenic Acid $pK_1 = 2.22$ $pK_2 = 6.98$ $pK_3 = 11.4$ Inorganic As V	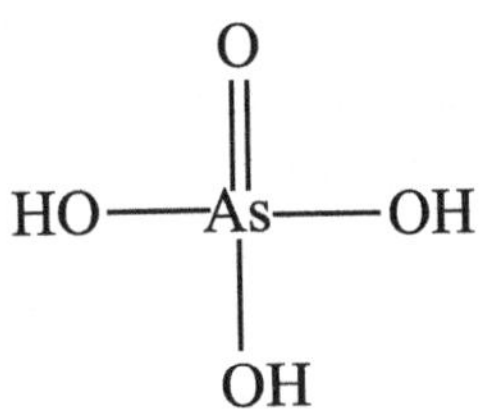	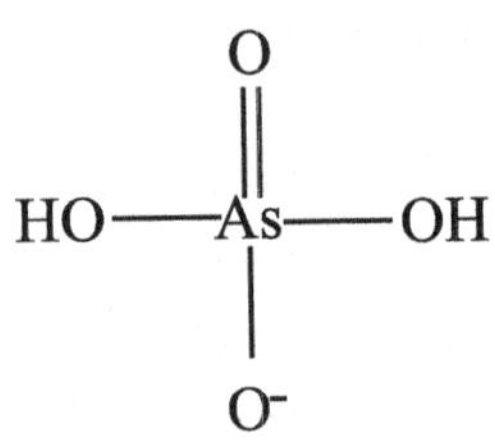 Monovalent anion	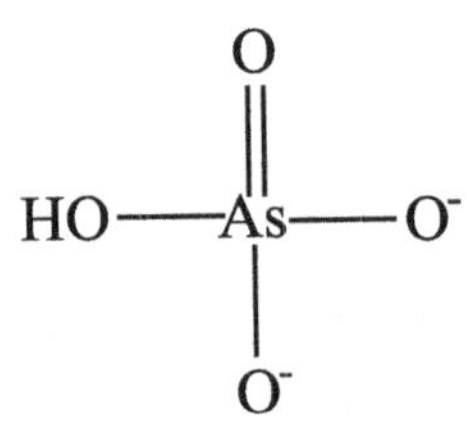 Divalent anion
Monomethylarsonic Acid—MMA $pK_1 = 3.41$ $pK_2 = 8.18$	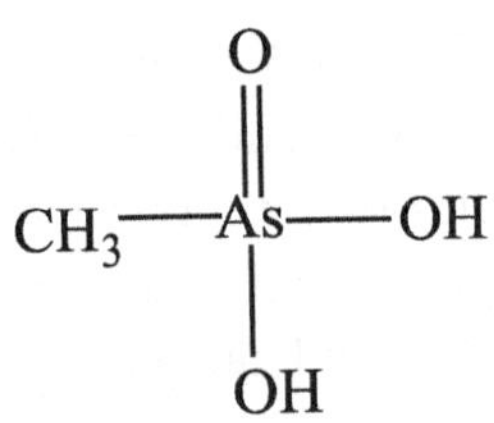	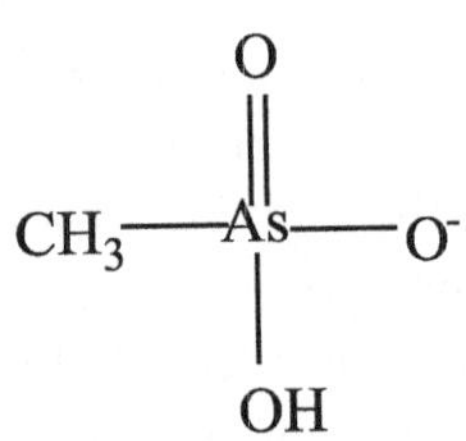 Monovalent anion	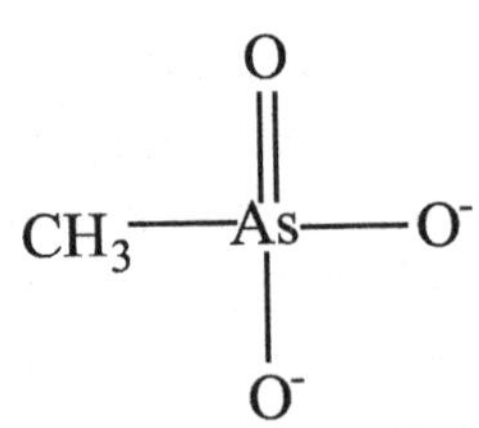 Divalent Anion
Dimethylarsinic Acid—DMA $pK_1 = 6.27$ Common name: Cacodylic acid	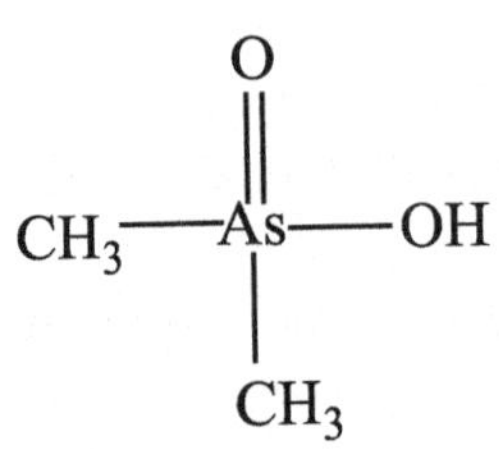	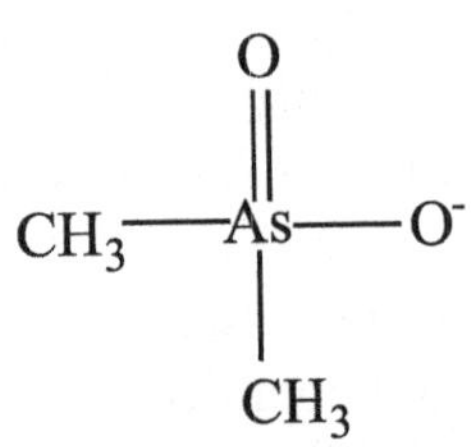 Monovalent anion	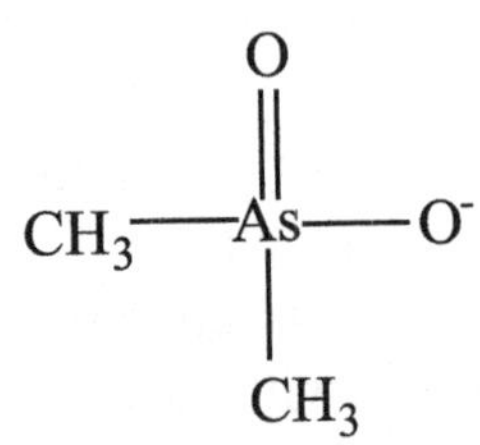 Monovalent anion

Table 1.2
Arsenic concentration and speciation in natural waters

Source	As (T) µg/L	As III %	As V %	MMA %	DMA %	Reference
Nova Scotia I	8000	56.3	43.8	<MDL	<MDL	Irgolic (1982)
Barefoot, AK	3100	77.4	22.6	<MDL	<MDL	Irgolic (1982)
Yenshei II, Taiwan	1100	2.2	98.2	<MDL	<MDL	Irgolic (1982)
Yenshei I, Taiwan	850	2.7	98.8	<MDL	<MDL	Irgolic (1982)
Antofagasta, Chile (Untreated)	750	2.1	98.7	<MDL	<MDL	Irgolic (1982)
Nova Scotia II	630	49.2	50.8	<MDL	<MDL	Irgolic (1982)
Antofagasta, Chile (Treated)	410	0.7	99.3	<MDL	<MDL	Irgolic (1982)
Hinkley, UT	180	5.6	94.6	<MDL	<MDL	Irgolic (1982)
Delta, UT	20	50	50	<MDL	<MDL	Irgolic (1982)
Hanford, CA (Well # 19)	90	100	0	ND	ND	Clifford (1999)
San Ysidro, NM (Well # 4)	250	100	0	ND	ND	Clifford and Lin (1991)
San Ysidro, NM (Well # 1)	88	35.2	64.8	ND	ND	Clifford and Lin (1991)
Hanford, CA (Various wells)	9-75	89	11	ND	ND	Hering and Chiu (2000)
Hanford, CA (Storage tank)	25	4-14	86-96	ND	ND	Hering and Chiu (2000)
Beaver Area, UT Well (M002)	30	77	23	<MDL	<MDL	Ficklin (1983)
Beaver Area, UT Well (M021)	3	0	100	<MDL	<MDL	Ficklin (1983)
Beaver Area, UT Well (M022)	7	29	71	<MDL	<MDL	Ficklin (1983)
New R. Braley, CA	11.3	0.1	99.5	0.1	0.1	Andreae (1977)
Squaw Lake, CA (Surface)	3.7	23.5	74.5	0.6	1.4	Andreae (1977)
Colorado R. Laguna Dam, CA	3.3	3.6	93.3	2.7	0.3	Andreae (1977)
Colorado R., Topcock, CA	2.8	3.1	81.1	4.7	11.2	Andreae (1977)
Kaweah R., Fish Camp, CA	2.8	3.8	88.7	0.3	7.2	Andreae (1977)
Coot Bay Pond, Everglades, FL	2.4	33.5	52.8	0.4	13.3	Andreae (1977)
Colorado R, Parker, AZ	2.2	5.2	89.5	2.9	2.3	Andreae (1977)

< MDL = Less than minimum detection level; ND = Not determined

(Andreae 1977; Bately and Low 1989). At natural water pH, both MMA and DMA exist predominantly as monovalent anions. At high pH values, MMA and DMA can be effectively removed from water, as divalent and monovalent anions, respectively, using hydroxide-form strong base anion (SBA) resin.

Arsenic speciation is also important because toxicity studies indicate that the inorganic forms-As III and As V-vary considerably in toxicity, and generally are much more toxic than the organic forms (NRC 1977).

NEED FOR FIELD ARSENIC SPECIATION METHOD

Due to redox reactions and photochemical oxidation, arsenic speciation of water samples changes over time (Gallagher et al. 2001). Often the biggest changes occur immediately following sampling and during shipping the samples from the field to the laboratory. Usually, As III is converted to As V, however the reduction of As V to As III has also been observed (Tong 1997). To ensure that the correct decision is made for arsenic treatment, As III/V speciation is required. Also, some health affect studies would benefit from knowledge of As III/V speciation. Therefore, arsenic-contaminated waters must be speciated immediately in the field or preserved for later speciation in the laboratory. Currently, there are no generally accepted (standard) methods for field speciation and/or preservation. In the light of the need for these methods, it was the objective of this project to develop a reliable field speciation method, which could eventually become a standard method.

ARSENIC SPECIATION METHODS

The inorganic and organic forms of arsenic expected in drinking water are readily separated by anion exchange. Of the various arsenic speciation methods available in the literature, the ones that demonstrated the greatest potential for adaptability for field use are listed in Table 1.3.

SPECIATION METHODS ADAPTABLE TO FIELD USE

Arsenic Speciation Methods of Clifford et al.

Clifford et al. (1983) developed an ion-exchange method using the chloride-form of commercially available ion-exchange resin. In the absence of iron (ferrous iron, Fe II), the sample pH may be in the range of 3-8. However, in the presence of Fe II and/or Fe III, As III passage and As V retention were adversely affected. Clifford recommended that if the presence of iron above about 0.3 mg/L was suspected, the pH of the water to be speciated be reduced to about 2.7-3.0 to avoid problems due to iron precipitation. When using the chloride-form resin, reducing the sample pH below 2.7 resulted in some As V leakage through the minicolumn, whereas, above pH 3, some As III was retained on precipitated $Fe(OH)_3$. Later the method was simplified and made faster by replacing the IX minicolumn with a 3M Empore ion-exchange filter disk (Ghurye and Clifford 2001).

Clifford's method differed from the previous methods (Ficklin 1983; Grabinski 1981) in that it used standard chloride-form ion-exchange resin beads (30-60 mesh). The larger beads allow for rapid gravity flow (1-10 mL/min) without much head loss and are much less expensive

Table 1.3
Arsenic speciation ion-exchange methods adaptable for field use

Ion-exchange methods for As speciation	As species determined	Type of resin	Procedure	Field use possible?
Grabinski (1981) Determination of As III, As V, MMA and DMA by IX chromatography with GFAAS detection	As III, As V, MMA, DMA	Cation: 20 mL SAC* 100-200 mesh H+ form Anion: 28 mL SBA† 100-200 chloride form	• Adjust pH to 2.5-7.0 range • <2 mL sample through resin • 55 mL 0.006 M trichloroacetic acid • 8 mL 0.2 M trichloroacetic acid • 55 mL 1.4 M NH_4OH • 50 mL 0.2 M trichloroacetic acid • GFAAS analysis of 3-20 mL fractions for separation • Approx. 30 min to separate all species using pressurized N_2	Method speciates inorganic and organic As. Not suitable for field use.
Ficklin (1983) Separation of As III and As V in ground water by ion-exchange.	As III, As V	Anion Resin 4 mL SBA†, 100-200 mesh, acetate form	• Acidify w HCl to pH <2 • 5 mL sample through resin column • Elute w 15 mL 0.12 M HCl • As III in first 10 mL • As V in last 10 mL	Yes, method developed for field use. Complicated HCl elution.
Clifford et al. (1983) As III/As V separation by Cl^- form Ion-exchange resins	As III, As V	Anion Resin 5 mL SBA† 30-60 mesh Chloride-form	• Adjust pH 3-8 range • 100 mL sample through resin • As III passes, As V retained • GFAAS analysis As (T), As III • As (T)- As III = As V • 30 mL of 0.5 M HCl for As V elution optional • Approx 10 min to speciate using gravity flow	Yes, method developed for both lab and field use. No elution of As V.
Edwards et al. (1998) Consideration in As analysis and speciation	As III, As V	Anion ~18 mL SBA† resin 50-100 mesh acetate form	• Acidify with H_2SO_4 to pH <2 • 40 mL sample through resin • As V by difference • ICP-MS analysis for total As • Approx 15 min to speciate using gravity flow	Yes, lab and field use. Modified Ficklin's method with H_2SO_4 and no elution of As V.
Ghurye and Clifford (2001)	Particulate As, As III, and As V	IX Filter Disk Anion Resin Filter Disk	• Adjust pH in 3-8 range • 20 mL sample through IXFD • As III passes, As V retained • AAS-Hydride Generation for As Analysis • MDL by above method established at 0.1 µg/L • Approx 2 min required to speciate	Yes, method developed for both lab and field use. Very fast speciation.

* SAC is typical polystyrene matrix strong acid cation resin
† SBA is typical polystyrene matrix strong base anion resin

than the finer (50-100 or 100-200 mesh) laboratory-grade resins. Furthermore, since anion resins are normally shipped in the chloride-form, conversion to acetate-form (via the hydroxide form) is not necessary. Despite the relatively large resin size, the results from the Clifford study indicated essentially complete separation of As III from As V. Finally, the Clifford method did not require pH adjustment nor did it require As V elution. Thus, the advantages of the Clifford method were simplicity and low cost, but at the risk of potential interference from iron.

Arsenic Speciation Methods of Ficklin and Edwards

Edwards et al. (1998) modified the Ficklin method (1983) for use in a national survey of arsenic in drinking water. Both methods use acetate-form anion-exchange resins and sample pH reduction to less than 1.5 (Ficklin) or 2.0 (Edwards). The significant differences between the methods are (a) the acidification procedure and (b) the lack of As V elution in the Edwards method. A minor difference is that the Edwards method uses a slightly coarser (50×100) mesh resin compared with Ficklin's method (100×200 mesh). Both mesh sizes are non-standard, however, and must be purchased from laboratory-supply houses. Edwards et al. (1998) used ICP-MS for arsenic analysis and preferred the acetate-form to the chloride-form for two reasons: (a) arsenic analysis by ICP-MS has major chloride interference, and (b) acetate is a base, which neutralizes the acid added to the sample and buffers the column effluent to near pH 4.75, the pK_a of acetic acid.

All three methods described above were chosen for further study because they speciate arsenic less than 15 minutes, are simple, and have the potential for development into a simple and robust field arsenic speciation method.

IMPORTANT FACTORS RELATED TO FIELD ADAPTABILITY

Resin Filter Disk vs. Minicolumn

Both resin filter disks and resin minicolumns were shown to be equally effective for arsenic speciation in the absence of iron in the water sample (Ghurye and Clifford 2001). Resin filter disks can be used quickly without extensive pretreatment, and a typical speciation can be performed in less than two minutes. A major drawback of filter disks is a much smaller ion-exchange capacity (approximately 0.2 meq/disk) compared with anion-exchange resins (10-14 meq per 10-mL resin minicolumn). Disks and minicolumns were extensively investigated here to determine their relative merits and drawbacks for field use.

Chloride vs. Acetate-Form of Resin

Both forms of resin, chloride and acetate, have been used by various researchers. As mentioned before, the Clifford method at pH 8.3 and 3.0 successfully used the chloride-form of both resin minicolumns and filter disks under a variety of experimental conditions (Clifford et al. 1983; Ghurye and Clifford 2001). Yet other researchers preferred the acetate-form of anion-exchange resins, mainly due to subsequent analytical considerations and the ability to lower sample pH to avoid iron precipitation. Because chloride is not a base, the sample pH must be about 3.0 or above when speciating arsenic using chloride-form resins.

Whereas there was no doubt that all three methods-Clifford with chloride-form resin, Clifford with IX-filter disks, and Edwards-would work equally well for iron-free arsenic-containing waters, their ability to overcome interference from the commonly occurring ferrous iron was not clear. This study focused on the efficiency of these methods in the presence of ferrous iron. Note that the Ficklin method is not discussed hereafter; for purposes of this study, the more recent Edwards method was used as representative acetate-form resin method.

Effect of Iron on Arsenic Speciation

In the presence of dissolved oxygen (DO) and above-neutral pH, dissolved iron (Fe (II)) will oxidize almost immediately and precipitate as iron hydroxide. It is known from previous research (Clifford et al. 1983; Edwards et al. 1998; Ghurye and Clifford 2001) that iron oxidation and precipitation introduces errors during inorganic arsenic speciation. Thus, the effect of iron oxidation and precipitation needed to be determined on the methods that would be considered for field use.

Effect of Sulfate on Arsenic Speciation

Based on the literature review, it was evident that feasible field speciation methods would involve the use of anion-exchange resins either in the chloride or acetate-form. With most resins, the concentration of sulfate determines the volume of sample that can be put through a fixed volume of resin before the capacity of the resin is exhausted. Therefore, the effect of sulfate concentration in the sample water on the retention of As V on the resin was determined by performing speciation tests with higher sulfate concentrations (50 mg/L to 240 mg/L).

Fate of Organic Arsenic in Inorganic Arsenic Speciation

The fate of organoarsenicals, MMA and DMA, during arsenic speciation is not known. If the organoarsenicals pass through the speciation media, they will be erroneously recorded as arsenite, whereas if the speciation media retains them, they will be erroneously recorded as arsenate. One of the goals of this study was to determine the fate of the organoarsenicals during inorganic arsenic speciation.

In general, the methods tested were evaluated on the basis of the criteria selection described in Table 1.4.

PROJECT OBJECTIVES

The overall objective of the arsenic speciation study was to develop a robust field arsenic speciation method to separate As III from As V for subsequent analysis of As III and As_T in an analytical laboratory. To achieve this objective following tasks were undertaken:

1. Identify existing speciation methods that show potential for adaptation to field use.
2. Compare the relative advantages and drawbacks of the identified speciation methods.
3. Evaluate the effect of ferrous iron on the identified speciation methods.
4. Select the best method and evaluate it for accuracy and robustness.
5. Determine the fate of organic arsenic (MMA and DMA) in the speciation method.
6. Develop a field protocol for the selected method, and verify the performance of the method in the field.

Table 1.4
Criteria for field method selection

Criterion	Remark
High As III/As V Selectivity	The candidate method must provide >90% recovery of As III and >90% retention of As V.
Short Time for Speciation	Less than 15 minutes
No pH Adjustment	pH adjustment to be avoided, if possible
Measure Particulate Arsenic	Method should attempt to differentiate between particulate and dissolved arsenic, if possible.
Minimal Interferences from (a) Particulate Iron (b) Sulfate (c) MMA and DMA	The optimized method should outline strategies for dealing with the said interferences.
Simplicity	Need for minimum or no specialized training for utilities personnel
Avoid Hazardous Chemicals	Avoid use of strong acids or bases if possible. Alternatively, investigate the use of milder acids and complexing agents.
Reuse Speciation Media	Speciation media should be regenerable and reusable, if possible.

CHAPTER 2
MATERIALS AND METHODS

REAGENTS AND STANDARDS

All chemicals were reagent grade and used without further purification. All reagent solutions were prepared with water purified using reverse osmosis and deionization using a Barnstead Nanopure system (here termed as DI water). All the glassware was cleaned by soaking in 10% HCl and rinsing four times with DI water.

In the experimental setup to obtain low DO water (described in Appendix B), the feed water was sparged with extra-dry-grade nitrogen gas resulting a residual dissolved oxygen (DO) level of approximately 0.05 mg/L. The arsenite stock solution (100 mg As/L) was prepared from the solid sodium m-arsenite FW 129.9, $NaAsO_2$,[1] dissolved in DI water. Ferrous iron stock solutions (1000 mg/L) were freshly prepared from $FeSO_4{\cdot}7H_2O$ every time before use in DO-free DI water. Reagent grade HCl or H_2SO_4 (1:1) was used to adjust the pH to 2.0 ± 0.1 and 3.0 ± 0.1 in both filter-disk and resin minicolumn experiments. EDTA solutions were prepared from the disodium salt of EDTA having MW of 378.02. Acetic acid (2.0 M), prepared from Glacial Acetic Acid (17.4 M), was used for pH adjustment in the EDTA-HAc experiments. The stock solution of ferrous sulfate[3] was used to spike the feed water for iron-contaminated experiments. The 200 mg/L stock solutions were prepared from both MMA[1] and DMA[1] (sodium salt of cacodylic acid), respectively and were used as the source of organic arsenic in the feed water.

Arsenic analysis was performed using a Perkin-Elmer 5000 Atomic Absorption spectrophotometer (AAS) at 193.7 nm and the hydride generation method using FIAS unit. Hydride generation was accomplished using analytical grade 0.2% (w/v) sodium borohydride pellets[2] dissolved in 0.05% sodium hydroxide.[3] Ultra-pure 0.14 M potassium iodide[2] was used as a prereductant for analysis of As V. Argon gas was used as the purge gas for the hydride generation unit. The standards for arsenic analysis were made from a 1000 mg/L As solution[1] (for both arsenic species). Four standards—1.3, 2.8, 6.8, 11.0 µg/L—and a preparation blank were prepared to establish a calibration curve, and the instrument was recalibrated every 20 samples. Organic arsenic (MMA and DMA) analyses were also performed using inductively coupled plasma mass spectrometry, using a Perkin Elmer Elan 6000 ICP-MS[4] equipped with a Gem-Tip cross-flow nebulizer, a Ryton spray chamber, a four-channel peristaltic pump and an autosampler. The standards used were 1, 2, 5, 10 µg-As/L. Iron analyses were performed using a Perkin Elmer[1] (AAnalyst 300) atomic absorption flame spectrophotometer with manual sample injection. Three standards—1, 2.5, 5.0 mg/L—and a blank were used to establish a calibration curve. Dissolved oxygen (DO) was measured in the deoxygenated samples using the Hach 8316 Indigo Carmine Method, which has a method detection limit of 0.01 mg/L.

---

1. Sigma Chemical Co, St. Louis, MO 63178
2. EM Science, Gibbstown, NJ 08027
3. Fisher Scientific, Fair Lawn, NJ 07410
4. Perkin Elmer Norwalk, CT

Table 2.1
Composition of NSF challenge water for arsenic speciation study

Cations	meq/L	mg/L	Anions	meq/L	mg/L
Na^+	3.197	73.5	HCO_3^-	3.0	183.0
Ca^{2+}	2.00	40.1	SO_4^{2-}	1.04	50.0
Mg^{2+}	1.04	12.6	Cl^-	2.0	71.0
			NO_3^--N	0.143	2.0
			F^-	0.053	1.0
			PO_4^{3-}-P	0.0013	0.04
			Silicate as SiO_2	0.33	19.8
Total	6.237	126.2		6.237	326.8

Calculated TDS = 466 – 93 mg/L (loss of H_2CO_3 during evaporation) = 373 mg/L
pH adjusted to the desired value using HCl or H_2SO_4

COMPOSITION OF CHALLENGE WATER

During the summer and fall of 2000, the authors had worked with the Arsenic Committee of the National Sanitation Foundation (NSF) to develop a universally-acceptable challenge water for testing arsenic-treatment devices. The Committee comprised twenty members representing NSF, water treatment equipment manufacturers and suppliers, government, and academia. Most importantly, this NSF Arsenic Challenge Water contains realistic concentrations of background contaminants such as silica, sulfate, phosphate, fluoride, and hardness, which are known to affect the arsenic capacity of adsorbents and ion-exchange resins. The USGS database for arsenic-contaminated potable ground waters was used to obtain the median concentrations of the background contaminants. A detailed recipe for the preparation of the NSF Arsenic Challenge Water was developed and tested at the University of Houston and at participating testing laboratories. The water was found to stable with respect to As III and hardness, i.e., As III did not oxidize to As V and $CaCO_3$ did not precipitate for at least 24 hours in the 8.2-8.5 pH range. Because of its acceptability and representative nature, this water was used as the basic test water for speciation testing. For the most of the experiments in this report, the pH of the synthetic test water was adjusted to 8.3 ± 0.1 using HCl. It was prepared by adding soluble salts to deionized water to obtain the desired water composition. A representative composition is shown in Table 2.1.

EXPERIMENTAL METHODS

Filter Disk Preparation

Commercially available IX filter disks[5] (having ion-exchange capacity of 0.2 meq/mL) were used as part of the initial speciation method screening studies, because they had been successfully used in previous research at the University of Houston (Ghurye and Clifford 2001).

Chloride-Form Filter Disk

The IX filter disk was first conditioned using organic solvents according to the manufacturers' recommendation (see Appendix A). Subsequent tests showed that organic-solvent conditioning had no effect on the arsenic throughput of the IX filter disks, thus it was discontinued. However, the IX filter disks, which were reportedly in the chloride-form as received, were treated to make sure they were in the chloride-form by sequentially passing 20 mL of 1 M HCl and 20 mL of 1 M NaCl, followed by a 20 mL-rinse using 0.005 M NaCl. Further details about IX filter disk activation and regeneration are provided in Appendix A.

Acetate-Form Filter Disk

In the initial experiments, the IX filter disks were put in the acetate-form by regenerating them with 20 mL of 1 M sodium acetate solution and then rinsing with 0.01 M sodium acetate solution. This regeneration was found to be insufficient for completely converting the IX filter disks to the acetate-form. Therefore, in subsequent experiments, the regenerant volume was doubled from 20 mL to 40 mL of 1 M sodium acetate. For detailed activation and regeneration procedures, see Appendix A.

Resin Preparation

Chloride-Form Resin Preparation

A commercially available SBA[6] resin was used to conduct the resin minicolumn experiments. The resin, supplied by the vendor in the chloride-form, was used directly without any pretreatment.

Acetate-Form Resin Preparation

A commercially available SBA resin[1] was used for acetate-form resin minicolumn experiments. As the vendor supplied the resin in the chloride-form, it was converted to the acetate-form according to the protocol of Edwards et al. (1998). Detailed procedures for resin conversion to the acetate-form are described in Appendix A.

---

5. Empore High Performance Extraction Disks, Filtration Products, St. Paul, MN
6. Dowex 1X8 50-100 Resin, Supelco, Bellefonte, PA

IX Filter Disk Experiments

One liter of NSF Arsenic Challenge Water was spiked with 0.5 mL of 100 µg/mL of As III stock solution to give a concentration of 50 µg/L As III in the feed water. Acceptable separation performance was defined as ≥90% As V retention and ≥90% As III passage. The As III feed water at a pH of 8.3 ±0.1 was passed through the filter disk in chloride- or acetate-form, and 5-mL aliquots were collected and analyzed for As III passage. Regeneration was performed after every 50 mL, and all the experiments were performed in triplicate. The As III feed water was then oxidized to convert As III to As V by dosing with chlorine stock solution to a level of 0.15 mg free Cl_2/L, i.e., three-times the stoichiometric requirement for complete oxidation of As III to As V (Ghurye and Clifford 2001). This As V feed water was then passed through the filter disk in chloride- or acetate-form and 5-mL aliquots were collected to determine As V retention.

Similar experiments were repeated at a pH of 2.0 ± 0.1 and 3.0 ± 0.1. Either concentrated HCl or 1:1 (v/v) sulfuric acid was used to lower the pH of feed water from 8.3 to 2.0 ± 0.1 and 3.0 ± 0.1. Samples for arsenic analysis were preserved by adding 5 µL of concentrated nitric acid to 5-mL aliquots to obtain a pH of less than 2.0.

Minicolumn Experiments

For the initial experiments, before optimization of resin volume, 5-mL (4-cm height) 0.7-cm ID Kontes Glass Flex columns were used to speciate As III and As V using 50-100 mesh size anion-exchange media. The anion-exchange resin was in the chloride-form (as stated by the vendor and it was used as received) or was converted to acetate-form as described in Appendix A. The minicolumn was filled with 5 mL of anion-exchange media, and the NSF Arsenic Challenge Water at pH 8.3 ± 0.1 was forced through it using a 60-mL syringe. The first 10 mL flush of the column effluent (speciated sample) was discarded, and the next 50 mL was collected. The speciated samples were then analyzed for As III passage or As V retention. The resin minicolumns were not regenerated, and a minicolumn with fresh resin was used each time, with each experimental condition performed in duplicate. The experiments were repeated at pH 2.0 ± 0.1 and 3.0 ± 0.1. Concentrated hydrochloric acid or 1:1 sulfuric acid was used to reduce the pH of NSF Arsenic Challenge Water. Acceptable performance was defined as ≥90% As V retention and as ≥90% As III passage. All samples taken for analysis were preserved by adding 5µL of nitric acid in the 5-mL aliquots to achieve a pH <2 in the preserved sample.

Preparation of DO-Free Challenge Water

In order to simulate anoxic groundwater and avoid the oxidation and precipitation of iron, Low DO challenge water was used for all the experiments that involved ferrous iron. Extra-dry grade nitrogen gas (<3 mg/L_v O_2) was sparged through the challenge water at pH 8.3. A schematic of the apparatus that was used to prepare DO-free challenge water and the other operational details are described in Appendix B. Sparging for one hour produced DO-free challenge water (DO~0.05 mg/L), and it was then spiked with 3.0 mg/L ferrous iron. After spiking with ferrous iron, the challenge water was further sparged for about 5 minutes and then the outlet valve was opened, and 500 mL of the arsenic-Fe II-spiked NSF Challenge Water was collected in a 500-mL bottle.

Clifford Method Procedure

From the 500-mL sampling bottle, 60-mL of arsenic- (As III or V) spiked-DO-free challenge water with 3 mg/L Fe II was then passed through a 5-mL minicolumn to speciate As III and As V using a 50-100-mesh size anion-exchange media in the chloride-form. The effluent from the minicolumn was analyzed for As III passage and As V retention. The same experiments were repeated at pH 2.0 ± 0.1 and 3.0 ± 0.1. pH reduction was achieved with either the use of conc. HCl or 1:1 (v/v) H_2SO_4. Once again, acceptable As III passage and As V retention were set at ≥90%.

Edwards Method Procedure

The Edwards method was performed in accordance with the recommended procedure described in a recent publication (Edwards et al. 1998). Polypropylene columns 100×15 mm (3.9×0.57 in.) were used by Edwards to speciate As III and As V using an acetate-form resin. DO-free challenge water was passed through the acetate-form of resin and the effluent was analyzed for As III passage and As V retention. Acceptable recoveries of As III were defined as ≥90% of As III in the effluent, and acceptable retention of As V was defined as ≥90% retention on the resin.

Clifford EDTA-HAc Method

A new method—the Clifford EDTA-HAc method based on the arsenic III/V preservation chemistry described by Gallagher et. al. (2001)—was developed at UH. It uses a complexing agent (EDTA) and a weak acid (HAc) to complex iron and reduce pH, respectively. DO-free challenge water was collected in 500-mL bottles, to which enough 0.01 M EDTA and 0.5 M HAc had been added to get a final concentration of 0.0005 M and 0.01 M, respectively. This stabilized NSF Challenge Water was then passed through a chloride-form resin minicolumn to determine As III recovery (passage) and As V retention.

CHAPTER 3
RESULTS AND DISCUSSION

The first section of this three-section chapter describes the results of preliminary studies, which examined the relative merits and drawbacks of the two speciation media examined: IX-filter disks and anion-exchange resin minicolumns. Both the chloride and acetate forms of the disks and minicolumns were evaluated. The preliminary studies were performed without Fe II in the challenge water. This important interference was studied in detail later.

The second section discusses the interference of iron oxidation during experiments performed using Fe II-spiked challenge water. After selecting resin media for further evaluation (IX filter disks were rejected based on the results of the preliminary studies), the performance of four arsenic speciation methods—the Clifford Method at pH 8.3, the Clifford Method at pH 3.0, the Edwards Method, and the EDTA-HAc Method—were compared in the presence of 3 mg/L Fe II.

The third section describes the results of the field experiments performed to develop the EDTA-HAc method into a robust and accurate field speciation method.

PRELIMINARY STUDIES WITHOUT FERROUS IRON (FE II)

Preliminary studies were performed using IX filter disks and resin minicolumns to determine the efficiency of both speciation media for inorganic arsenic separation. These experiments were conducted at pH values of 8.3 ± 0.1, 2.0 ± 0.1, and 3.0 ± 0.1, using both chloride- and acetate-form filter disks and resins. Either concentrated hydrochloric acid or 1:1 (v/v) sulfuric acid was used to reduce the pH of feed water for low-pH (pH 2.0 and 3.0) experiments. The speciated samples were analyzed for As III passage or As V retention. For this study, acceptable performance was defined as ≥90 % As III passage and ≥90 % As V retention.

Chloride-Form IX Filter Disks

Chloride-form IX filter disks were successfully used in a previous study that examined various techniques for As III oxidation (Ghurye and Clifford 2001). A key requirement in the As III oxidation study was the need to quantitatively separate As III and As V within 10 seconds under a variety of experimental conditions. The chloride-form IX filter disks are easily prepared and regenerated, and can easily accommodate flow rates as high as 60 mL/min. One problem of the IX filter disk is its low ion-exchange capacity, 0.2 meq/disk, compared with 10-14 meq/minicolumn for separating columns containing 10 mL of resin. Thus, there is a fundamental capacity limitation on the size of a sample that can be speciated with an IX filter disk. We considered the use of multiple IX disks to increase the exchange capacity but rejected the idea because of cost, approximately $8 per disk.

Necessity for Organic-Solvent Conditioning

As described in detail in the Materials and Methods chapter, the manufacturer recommended that the IX filter disks be conditioned with organic solvents for optimum performance. However, the recommended conditioning failed to meet two basic criteria of the new method: simplicity and avoidance of hazardous chemical usage. Therefore, the performance

(As III passage and As V retention) of an IX filter disk that had been simply wetted by passing 20 mL of 0.005 N NaCl through it was compared with the performance of an organic solvent-conditioned filter. The results of organic conditioning followed by HCl-NaCl pretreatment can be seen in Figures 3.1 and 3.2 for As III and As V, respectively. Figures 3.3 and 3.4 show the As III and As V performance of the IX filter disks without organic conditioning but with HCl-NaCl pretreatment. Finally, the performance of minimally-pretreated (by simply wetting with 20 mL of 0.005 NaCl) IX filter disks is shown in Figures 3.5 and 3.6 for As III and As V, respectively. As seen in Figures 3.1 through 3.6, simply rinsing the filter with 0.005 M NaCl gave similar speciation performance as an organic solvent-NaCl/HCl conditioned filter and a NaCl/HCl conditioned filter. After initial use (passage of 50 mL), the minimally-pretreated IX filter disks were regenerated using 20 mL of 1 M HCl, 20 mL of 1 M NaCl, and a 20-mL 0.005 M NaCl rinse. The performance of the regenerated IX filter disk was similar to the performance of a fresh IX filter disk. Thus, preliminary studies indicated that, in the absence of ferrous iron, the IX filter disks in the chloride-form can be used for efficient As III/As V separation. The IX filter disks can be used with minimal pretreatment and can be efficiently regenerated.

Subsequent experiments with chloride-form IX filter disks were performed with fresh and regenerated filters where the IX filter disk, prior to initial use, had been minimally pretreated by wetting with 20 mL of 0.005 M NaCl only.

At pH 3.0, which used in the 1983-Clifford minicolumn method for sample containing iron, quantitative As III passage was obtained (Figure 3.7) but virtually no As V was retained on the chloride-form IX filter disk (Figure 3.8) due to very low capacity (meq of As V per meq of disks capacity) of the filter disks at low pH. This suggests that if the pH of the feed water must be reduced to prevent iron precipitation, the chloride-form IX filter disks will not be able to retain As V.

Further experiments using HCl or H_2SO_4 for acidification were performed at pH 2.0, which is the operating pH for the Edwards Method. Once again, as seen from Figures 3.9 through 3.12, at pH 2.0, using either HCl or H_2SO_4, As III was quantitatively recovered in the IX filter disk effluent, i.e., acceptable As III recovery was achieved. However, As V retention was minimal in all cases because As V is present predominantly as fully protonated species. Thus, the chloride-form IX filter disks failed to retain As V in all the low-pH experiments, which ruled out their use for field speciation.

It was further observed that As III recoveries did not reach 100% until at least 15 mL of feed water had passed through the IX filter disks. The reason for this low As III recovery in the first 15 mL is that the filter holder has approximately 5-7 mL of displacement volume that is filled with water before speciation. This displacement water dilutes the influent coming into the filter holder, thus causing the effluent As III concentration to be less than the influent concentration of 50 μg/L.

Acetate-Form IX Filter Disks

Acetate-form IX filter disks were also evaluated at pH 8.3, 3.0 and 2.0. Initially, the filter disks were converted to the acetate form by passing 20 mL of 1 M sodium acetate (20 meq sodium acetate). It was assumed that the 100-fold excess of acetate ions (20 meq acetate/0.2 meq disk capacity) would be more than enough to convert the disks from the chloride form to the acetate form. When challenge water was passed through the IX filter disk, acceptable As III passage was obtained after the first 15 mL of filtrate (Figure 3.13). Ignoring the first 15 mL of As V filtrate

(Figure 3.14), we see that the acetate-form IX filter disks were incapable of good As V retention beyond 10-15 mL of additional As V sample throughput. However, when the initial regenerant volume was doubled to 40 mL (40 meq Ac^-, a 200-fold excess) when putting the disks in the acetate form, As III passage and As V retention were within acceptable (90%) range. See Figures 3.15 and 3.16.

At the lower pHs studied using the acetate-form IX filter disks, the results were qualitatively similar to those obtained with the chloride-form IX filter disks. As III passed through the disk at both pH 2.0 and 3.0, which is not surprising as it is fully-protonated at these pH values (pK_a = 9.22). Although, As V retention was somewhat higher at pH 3 than at pH 2, it was not enough to be termed successful. Pertinent figures are included in Appendix C. Table 3.1 provides a summary of the IX-filter-disks (chloride- and acetate-form) experiments performed as a part of preliminary studies.

To summarize, the chloride-form IX filter disks provided acceptable performance only at pH 8.3 and showed no significant As V capacity at pH 2-3. Similar results were obtained using the acetate-form IX filter disks where acceptable performance was obtained at pH 8.3 but, again, the method failed at pH 2-3.

RESIN MINICOLUMN EXPERIMENTS

As seen from the results of the IX filter disk experiments, one of the important drawbacks of the disks is their limited ion exchange capacity. This capacity restriction can be overcome by using resin in minicolumns. Experiments performed using resin minicolumns were similar to the IX filter disk experiments. Briefly, both chloride- and acetate-forms of resin were studied at pHs of 8.3, 3.0 and 2.0. pH reduction was performed using either HCl or H_2SO_4.

Chloride-Form Resin Minicolumn Experiments

Initially, 7-mL resin minicolumns were used to conduct the preliminary experiments. Note that the resin volume was increased to 10 mL during the method development studies. As-received chloride-form resin[7] was used directly without any pretreatment.

When using the arsenic challenge water at pH 8.3, the IX minicolumn produced acceptable results for As III passage and As V retention (Figures 3.17 and 3.18). At pH 3.0, acceptable As III and As V performances were obtained when HCl was used for pH adjustment (Figures 3.19 and 3.20). The successful performance at pH 3.0 is supported by the results of the Clifford Method (Clifford et al. 1983), which recommended an optimum low-pH range of 2.7-3.1 for As III/As V separation. However, when H_2SO_4 was used for pH adjustment, the 7-mL Cl-form resin minicolumn essentially had no capacity for As V retention (see Figures 3.21 and 3.22). This result was not unexpected: sulfate and bisulfate are preferred ions on conventional SBA resins, and their concentrations resulting from sulfuric acid usage essentially overwhelmed the As V retention capacity of the resin.

As expected at pH 2.0, acceptable As III passage was obtained using either HCl or H_2SO_4 for pH adjustment because As III exists primarily as fully-protonated uncharged arsenious acid (H_3AsO_3) below pH 8.2 (one pH unit less than its pK_a of 9.2). In a similar manner, As V, which

---

7. SBA resin, IX8, Supelco-Aldrich, $168/500g

Table 3.1
Summary of the preliminary IX filter disk experiments without Fe II present

No.	IX filter	pH	Acid	Feed	Result	Comment
1	Chloride*	8.3	—	As III	Success	>90% As III passage after 15 mL
2	Chloride*	8.3	—	As V	Success	Acceptable As V capacity
3	Chloride†	8.3	—	As III	Success	>90% Passage of As III after 15 mL
4	Chloride†	8.3	—	As V	Success	Acceptable As V capacity
5	Chloride	8.3	—	As III	Success	>90% As III passage after 15 mL
6	Chloride	8.3	—	As V	Success	Acceptable As V capacity
7	Acetate‡	8.3	—	As III	Success	>90% As III passage after 15 mL
8	Acetate‡	8.3	—	As V	Failure	Breakthrough in As V after 35 mL
9	Acetate§	8.3	—	As III	Success	>90% As III passage after 15 mL
10	Acetate§	8.3	—	As V	Success	Acceptable As V retention
11	Chloride	2.0	H_2SO_4	As III	Success	>90% As III passage after 15 mL
12	Chloride	2.0	H_2SO_4	As V	Failure	No capacity for As V retention
13	Chloride	2.0	HCl	As III	Success	>90% As III passage after 25 mL
14	Chloride	2.0	HCl	As V	Failure	No capacity for As V retention
15	Acetate	2.0	H_2SO_4	As III	Success	>90% As III passage after 15 mL
16	Acetate	2.0	H_2SO_4	As V	Failure	No significant capacity for As V
17	Acetate	2.0	HCl	As III	Success	>90% As III passage after 20 mL
18	Acetate	2.0	HCl	As V	Failure	No significant capacity for As V
19	Chloride	3.0	HCl	As III	Success	>90% As III passage after 15 mL
20	Chloride	3.0	HCl	As V	Failure	No significant capacity for As V
21	Acetate	3.0	HCl	As III	Success	>90% As III passage after 15 mL
22	Acetate	3.0	HCl	As V	Failure	No significant capacity for As V

* Fresh filter wetted with 0.005 M NaCl;
† Organic Conditioned Filter;
‡ insufficient 20-x regenerant: 20 mL 1 M NaAc and 20 mL 0.01 M NaAc;
§ sufficient 40-x stoichiometric regeneration
Regeneration volumes unless otherwise specified—For Chloride form of filter: 20 mL 1 M HCl, 20 mL 1 M NaCl, 20 mL 0.005 M NaCl; For Acetate Form of Filter: 40 mL 1 M NaAc, 40 mL of 0.01 M NaAc

has a pK_a of 2.2 and is mostly protonated (uncharged) at a pH 2.0 is poorly retained on the resin, which caused As V separation to fail. Pertinent figures are included in Appendix D.

Acetate-Form Resin Minicolumn Experiments

The SBA resin was converted to the acetate-form according to the procedure recommended by Edwards et al. (1998). Relevant details are included in Material and Methods and in Appendix A. When arsenic challenge water with 50 µg/L As III at pH 8.3 was passed through the acetate-form resin minicolumn, greater than 90% passage of As III was obtained (Figure 3.23). Acceptable retention of 50 µg/L As V was also obtained at pH 8.3 (Figure 3.24). Similar results were obtained at pH 3.0 when either HCl or H_2SO_4 was used for pH adjustment (See Figures 3.25-3.28).

As expected at pH 2.0, acceptable (≥90%) As III passage was obtained using either HCl or H_2SO_4 for pH adjustment (See Figures 3.29 and 3.31). However, although As V was retained on the resin, the throughput before significant As V breakthrough was only 30 mL (with both HCl and H_2SO_4 for pH adjustment) (See Figures 3.30 and 3.32). The failure to retain As V for more than 50 mL at pH 2.0 seems primarily to be an exchange capacity issue. That is, acceptable performance can be obtained by increasing resin volume from 7 mL to 18.5 mL, which is the volume or acetate-form resin that Edwards et al. (1998) used to provide sufficient capacity for As V at a feed pH near 2.0 using sulfuric acid.

The role of buffering that acetate plays deserves some mention. One reason that the acetate-form resin was partially successful for As V retention, whereas the chloride-form proved completely inadequate, is the buffering capacity provided by the acetate-form of the SBA resin. When using the acetate-form resin, the pH of effluent was experimentally measured at about 3.8, which is higher than the pH-2.0 water that went into the resin minicolumn. The higher effluent pH is due to the buffering capacity of the acetate-form resin. At pH 3.8, As V is ionized and is thus retained on the resin as $HasO_4^-$ as opposed to its complete passage as un-ionized H_3AsO_4 through the un-buffered chloride-form resin at pH 2.0. The buffering provided by the acetate-form resin when the influent is acidified to a low value is an important advantage of the Ficklin and Edwards methods. The inadequate throughput from 7-mL resin volume is addressed later in the Methods Development section.

Table 3.2 lists all the resin minicolumn experiments performed during preliminary studies.

To summarize, the chloride-form resin minicolumn provided acceptable speciation performance at pH 8.3 and 3.0 in the presence of HCl only. It failed at pH 2.0 in the presence of both HCl and H_2SO_4 because As V at pH 2.0, about half of the As V is fully protonated, i.e., exists as an undissociated molecule. The acetate-form minicolumn also provided acceptable speciation performance at pH 8.3 and 3.0 in the presence of both HCl and H_2SO_4. It partially failed at pH 2.0 because of insufficient resin volume, 7 mL, to retain As V. This situation is easily corrected by adding more resin to the column, so it is not a true failure.

SPECIATION EXPERIMENTS WITH FERROUS IRON

In the presence of dissolved oxygen (DO) above neutral pH, dissolved ferrous iron will be oxidized almost immediately and precipitate to form ferric hydroxide $Fe(OH)_3$ or ferric oxyhydroxide (FeOOH), adsorb As V (and to a lesser extent As III) and thus interfere with arsenic

Table 3.2
Summary of results for resin minicolumn experiments without Fe II present

No.	Resin	pH	Acid	Feed	Result	Comment
1	Chloride	8.3	—	As III	Success	>90% As III passage after first 10 mL
2	Chloride	8.3	—	As V	Success	Acceptable As V capacity
3	Chloride	2.0	HCl	As III	Success	>90% As III passage after first 10 mL
4	Chloride	2.0	HCl	As V	Failure	No significant capacity for As V
5	Chloride	2.0	H_2SO_4	As III	Success	>90% As III passage after first 10 mL
6	Chloride	2.0	H_2SO_4	As V	Failure	No capacity for As V retention
7	Chloride	3.0	HCl	As III	Success	>90% As III passage after first 10 mL
8	Chloride	3.0	HCl	As V	Success	Acceptable As V retention
9	Chloride	3.0	H_2SO_4	As III	Success	>90% As III passage after first 10 mL
10	Chloride	3.0	H_2SO_4	As V	Failure	No capacity for As V
11	Acetate	8.3	—	As III	Success	>90% As III passage after first 10 mL
12	Acetate	8.3	—	As V	Success	Acceptable As V capacity
13	Acetate	2.0	HCl	As III	Success	>90% As III passage after first 10 mL
14	Acetate	2.0	HCl	As V	Failure	Breakthrough after 30 mL*
15	Acetate	2.0	H_2SO_4	As III	Success	>90% As III passage after first 10 mL
16	Acetate	2.0	H_2SO_4	As V	Failure	Breakthrough after 30 mL*
17	Acetate	3.0	HCl	As III	Success	>90% As III passage after first 10 mL
18	Acetate	3.0	HCl	As V	Success	Acceptable As V capacity
19	Acetate	3.0	H_2SO_4	As III	Success	>90% As III passage after first 10 mL
20	Acetate	3.0	H_2SO_4	As V	Success	Acceptable As V capacity

* Early breakthrough of As V was due to insufficient volume, 7 mL, of resin

speciation. In order to minimize the interference of dissolved iron, DO-free challenge water was produced using the experimental set up and apparatus described in Chapter 2.

Four potential field methods were evaluated for waters containing 3 mg/L ferrous iron. These were:

1. Clifford Method at pH 8.3 using chloride-form resin
2. Clifford Method at pH 3.0 using HCl acidification and chloride-form resin
3. Edwards Method at pH 2.0 using H_2SO_4 acidification and acetate-form resin
4. EDTA-HAc Method at pH 3.6 using chloride-form resin

Clifford Method at pH 8.3

DO-free arsenic-spiked challenge water was used to test the 1983 Clifford Method at pH 8.3. The challenge water (batch of 4.0 L) for the experiments was prepared as explained in Chapter 2.

When DO-free As III-spiked challenge water containing ferrous iron was passed through the chloride-form of resin at pH 8.3, less than 90% passage of As III was obtained. At pH 8.3, the oxidation rate of ferrous in the presence of air is fast. Once ferric hydroxides are formed, the oxidation of As III to As V takes place rapidly (Hug et al. 2001). So during speciation experiments with As III in the presence of Fe II, less than 90% passage of As III was obtained due to the oxidation of As III. When As V was used in the Fe II-containing challenge water, a significant amount of As V was obtained in the effluent. This As V was likely adsorbed onto ferric hydroxides, which were formed at pH 8.3 and passed through the speciating column as particles. Thus, at pH 8.3 with 3 mg/L Fe II present, the Clifford Method failed to produce acceptable results due to iron precipitation.

Clifford Method at pH 3.0

The Clifford method for iron-containing waters is the pH 3.0, chloride-form resin minicolumn method. DO-free NSF Challenge Water was spiked with 3 mg/L Fe II, and then a sample was collected in a beaker. The pH was reduced to 3.0 using HCl solution and then speciated for As III passage and As V retention. The method gave barely 90% As III recovery even after 20 mL. Because Fe II-containing sample was exposed to air for 5-6 min before pH adjustment and speciation and during this period Fe II was oxidized and precipitated as FeOOH. Also, some As III was possibly oxidized to As(V) in the presence of Fe II and ultraviolet (room) light (Hug et al. 2001). As V leakage was observed, presumably because of adsorption of As V onto iron precipitate and subsequent leakage through the speciation column. See Figures 3.33 and 3.34.

The reason for the failure of the Clifford Method at pH 3.0 is that some iron precipitation was observed even before the pH of the feed water was adjusted to 3.0. Even after acidification to pH 3.0, this precipitate did not fully redissolve, and some As V, adsorbed onto the precipitated iron, leaked through the resin minicolumn. The most important conclusion from the pH 3.0-Clifford Method test was that iron precipitation prior to speciation would prevent accurate arsenic III/V speciation.

Edwards Method at pH 2.0

In order to carefully replicate the Edwards Method, field speciation kits (provided by Edwards) were used to perform As III/As V speciation at pH 2.0 in strict accordance with his recommended procedure. Sulfuric acid instead of HCl is used in the Edwards method for acidification to pH 2.0 to eliminate chloride interference during subsequent arsenic analysis using ICP-MS.

Speciation was performed by two different operators[8] according to the Edwards method, which begins with a 0.45 μm filtration step followed by acidification of the filtrate with H_2SO_4 and

---

8. Saqib Karori, MS Student and Ganesh Ghurye, Ph.D. student, Department of Civil and Environmental Engineering, University of Houston, Houston TX

subsequent speciation using acetate-form resin. The flow rate from the 4-L nitrogen-sparging bottle was adjusted to 100 mL/min and a flowing sample was collected in a 1-L beaker. Major iron precipitation was visually observed to occur during the time a 1-L challenge water sample was collected in the beaker. Figure 3.35 shows the As III passage through the acetate-form of resin and Figure 3.36 shows the As V retention on the acetate-form resin using the Edwards Method at pH 2.0. The pH of the effluent coming out of the acetate-form resin was 4.1.

It appeared that ferrous iron added to the challenge water completely precipitated before acid was added, because no significant iron was found in the filtrate. Most As V and a significant fraction of As III was adsorbed on the ferric hydroxide precipitate. The precipitated ferric hydroxide with some adsorbed arsenic was removed during the initial filtration step so that the filtrate contained only a fraction of As V or As III present in the original feed water. Thus, the presence of ferrous iron will result in the mis-identitification of once-soluble arsenic as particulate arsenic. Approximately 15 μg/L of the As III and 33 μg/L of As V (from an initial As III or As V concentration of 50 μg/L) were removed during the filtration step. Once acidified to pH <2, the Edwards method successfully allowed the passage of the 35 μg/L As III that remained in solution after filtration. This was unlike the Clifford method at pH 3.0, which did not have a filtration step, and, at pH 3.0, allowed most of the precipitated iron with adsorbed As III or As V to pass through the ion-exchange column unhindered (See again Figures 3.33 and 3.34). Approximately 17 μg/L As V of original 50 μg/L that remained in solution was quantitatively retained by the acetate-form resin during speciation. Nevertheless, the Edwards method failed to meet the requirements of an accurate and robust filed speciation method because it failed to overcome interference due to ferrous iron during the filtration step. Our prediction for this method when used on waters containing As III and ferrous iron is that it is likely to falsely report some soluble As III as particulate arsenic while understating the soluble As III concentration.

EDTA-HAc Method

A new method, which uses acetic acid (HAc) to reduce pH to approximately 3.5 and EDTA to complex iron, was developed. It is based on the arsenic species preservation chemistry developed by USEPA (Gallagher et al. 2001). It was observed in the other methods that it was difficult to eliminate the effect of iron during the arsenic speciation, even when the feed pH was reduced to as low as 2.0. This brought to fore the necessity of immediately complexing iron prior to arsenic speciation. EDTA is a well-known chelating agent and rapidly complexes both ferrous and ferric ions in solution. If the initial 15-mL column-flushing volume of the speciated sample is ignored (especially in case of As III), then the method gives acceptable As III passage and As V retention as shown in Figures 3.37 and 3.38 respectively.

Effect of Added Reagents on Arsenic Analysis Using HGAA

Prior to conducting EDTA-HAc Experiments, the potential effect of these added reagents on arsenic analysis using HGAA was studied. Acetic acid and EDTA were added to arsenic-containing challenge water to achieve a final concentration of 0.01 M and 0.0005 M respectively. This concentration of EDTA was chosen because it is capable of complexing 0.0005 M Fe, i.e., up to 27 mg/L Fe II or Fe III in the absence of mg/L levels of other strongly complexed metals such as lead and copper. Calcium and magnesium were not considered as serious competitors for EDTA binding because their stability constants are 3-5 orders of magnitude less than Fe II and

13-16 orders of magnitude less than Fe III (*Lange's Handbook,* 14th ed.) The added reagents, viz., HAc, EDTA and Fe II were found to have no significant effect on arsenic analysis by HG-AAS. Relevant details are included in Appendix E.

Out of four methods extensively studied, it was found that only the EDTA-HAc method was capable of accurate As III/V field speciation in the presence of 3 mg/L dissolved iron. In fairness, it must be noted that the Edwards method failed primarily because of its mandatory filtration step, which was not included in the other speciation methods by design.

COMPARISON OF ALL FOUR SPECIATION METHODS

Four speciation methods—the pH 8.3 Clifford Method, the pH 3.0 Clifford Method, the Edwards Method, and the EDTA-HAc Method—were used to speciate solutions containing 2-3 mg/L Fe II and and 50 µg/L As III or As V. The fate of iron and either As III or As V during the course of the speciation process is described below. To better understand the performance of each of the speciation methods, the fate of iron for these four methods is presented first, followed by the arsenic speciation efficiency.

Fate of Ferrous Iron vs. Speciation Method

Tables 3.3 and 3.4 show the results of the iron experiments for challenge waters containing ferrous iron and 50 µg/L As III or 50 µg/L As V. With the notable exception of the EDTA-HAc Method, the other three methods (pH 8.3-Clifford, pH 3.0-Clifford and pH 2.0-Edwards) were unable to overcome interference due to precipitated iron. Of these three methods, the one most affected by iron precipitation was the Edwards method. This was primarily due to the filtration step in the Edwards method, which requires a 1-L sample to be collected in an open 1-L beaker, before speciation can proceed. The EDTA-HAc Method, on the other hand, successfully kept all of the iron in solution because samples were taken directly into bottles containing the EDTA-HAc reagents.

Efficiency of Four Arsenic Speciation Methods in the Presence of Ferrous Iron

The four (above mentioned) methods were evaluated for arsenic speciation efficiency in the presence of ferrous iron, and the results for As III and As V are presented in Tables 3.5 and 3.6, respectively. As seen from these Tables 3.5 and 3.6, the speciation accuracy was directly related to the ability of the method to keep iron in solution. Not surprisingly, the Edwards method, which allowed oxidation and precipitation of nearly all the iron even before the filtration step, arsenic speciation efficiency was the poorest. It gave the highest values for particulate arsenic due to the near-complete precipitation of iron. However, as noted before, once sulfuric acid was added and the pH lowered to 2.0, the Edwards method successfully retained As V and allowed the quantitative passage of unadsorbed As III through the acetate-form resin.

Figures 3.39 and 3.40 compare the four methods for percent As III recoveries and percent As V retention. From these figures, it can be seen that EDTA-HAc Method gives acceptable (>90%) As III recoveries after initial 15 mL whereas Clifford Method at pH 3.0 and Clifford Method at pH 8.3 fails to recover As III and also to retain As V. However, Edwards method gives acceptable As V retention, but fails to recover acceptable As III in the effluent from the acetate-form resin minicolumn.

Table 3.3
Comparison of soluble and particulate Fe in test waters with As III

Method	Pretreatment	Feed Fe (mg/L)	Soluble Fe in 0.2 μm filtrate (mg/L)	Soluble Fe in 0.45 μm filtrate (mg/L)	Particulate Fe by difference (mg/L)*	Air contact time (minutes)†
EDTA-HAc Method	Take sample in bottle with HAc and EDTA	2.51	2.54	N/A	0.0	0
		2.51	2.52	N/A	0.0	0
		2.52	2.51	N/A	0.0	0
Clifford Method pH 3.0	Reduce pH using HCl in an open beaker	2.33	1.78	N/A	0.5	5 – 6
		2.35	1.74	N/A	0.6	5 – 6
		2.37	1.74	N/A	0.6	5 – 6
Clifford Method pH 8.3	None	2.84	1.28	N/A	1.6	1 – 3
		2.92	1.30	N/A	1.6	1 – 3
		2.95	1.30	N/A	1.7	1 – 3
Edwards Method	Reduce pH by 0.05% H_2SO_4 after filtration	2.77	0.14	0.27	2.6	5 – 10
		2.97	0.14	0.25	2.8	5 – 10
		2.76	0.12	0.2	2.6	5 – 10
		2.76	0.10	0.2	2.7	5 – 10

N/A- Not analyzed

* Particulate Fe is the difference between "Feed Fe" and "soluble Fe in 0.2 μm filtrate" i.e., Column 3 minus Column 4.

† Air contact time is the time of sample exposed to air before preservation, or filtration, or the actual speciation in the minicolumn. Iron precipitation in the Clifford method at pH 3.0 and 8.3 indicates that oxidation occurs during speciation in the minicolumn.

Table 3.4
Comparison of soluble and particulate Fe in test waters with As V

Method	Pretreatment	Feed Fe (mg/L)	Soluble Fe in 0.2 μm filtrate (mg/L)	Soluble Fe in 0.45 μm filtrate (mg/L)	Particulate Fe by difference (mg/L)*	Air contact time (minutes)†
EDTA-HAc Method	Take sample in bottle containing HAc and EDTA	2.81	2.81	N/A	0.0	0
		2.75	2.80	N/A	0.0	0
		2.78	2.82	N/A	0.0	0
Clifford pH 3.0	Reduce pH by conc. HCl in an open beaker	2.92	1.51	N/A	1.4	5 – 6
		2.92	1.51	N/A	1.4	5 – 6
		2.93	1.51	N/A	1.4	5 – 6
Clifford pH 8.3	None	2.84	1.28	N/A	1.6	1 – 3
		2.92	1.30	N/A	1.6	1 – 3
		2.95	1.30	N/A	1.7	1 – 3
Edwards Method	Reduce pH by 0.05% H_2SO_4 after filtration	2.88	0.07	0.08	2.8	5 – 10
		2.89	0.07	0.09	2.8	5 – 10
		2.74	0.05	0.07	2.7	5 – 10
		2.75	0.05	0.04	2.7	5 – 10

N/A- Not analyzed

* Particulate Fe is the difference between "Feed Fe" and "soluble Fe in 0.2 μm filtrate" i.e., Column 3 minus Column 4.

†Air contact time is the time of sample exposed to air before preservation or filtration or the actual speciation in the minicolumn. Iron precipitation in the Clifford method at pH 3.0 and 8.3 indicates that oxidation occurs during speciation in the minicolumn.

Table 3.5
Comparison of the dissolved and particulate As III in all the four methods studied for potential use for field application in waters containing iron

Method	Pretreatment	Feed As III (μg/L)	Soluble As III in 0.2 μm filtrate (μg/L)	Soluble As III in 0.45 μm filtrate (μg/L)	Particulate As III by difference (μg/L)*	Air contact time (minutes)†
EDTA-HAc Method	Take sample in bottle containing HAc and EDTA	44.6	44.3	N/A	0.2	0
		44.5	45.0	N/A	-0.5	0
		44.5	45.7	N/A	-1.2	0
		43.6	35.2	N/A	8.4	0
Clifford's pH 3.0	Reduce pH by conc. HCl in an open beaker	43.3	34.7	N/A	8.6	5 – 6
		43.6	35.2	N/A	8.4	5 – 6
		53.4	45.7	N/A	7.7	5 – 6
Clifford's pH 8.3	None	51.0	45.7	N/A	5.4	1 – 3
		52.9	45.7	N/A	7.2	1 – 3
		47.9	32.3	N/A	15.6	1 – 3
Edwards Method	Reduce pH by 0.05% H_2SO_4 after filtration	47.2	32.1	32.1	15.1	5 – 10
		46.2	33.8	35.7	12.4	5 – 10
		47.2	33.0	36.4	14.3	5 – 10

N/A- Not analyzed
* Particulate As III is the difference between "Feed As III" and "soluble As III in 0.2 μm filtrate" i.e., Column 3 minus Column 4.
† Air contact time is the time of sample exposed to air before preservation or filtration or the actual speciation in the minicolumn. Iron precipitation in the Clifford method at pH 3.0 and 8.3 indicates that oxidation occurs during speciation in the minicolumn.

Table 3.6
Comparison of the dissolved and particulate As V in all the four methods studied for potential use for field application in water containing iron

Method	Pretreatment	Feed As V (µg/L)	Soluble As V in 0.2 µm filtrate (µg/L)	Soluble As V in 0.45 µm filtrate (µg/L)	Particulate As V by difference (µg/L)*	Air contact time (minutes)†
EDTA-HAc Method	Take sample in bottle containing HAc and EDTA	48.2	45.9	N/A	2.3	0
		48.0	47.7	N/A	0.3	0
		46.0	45.5	N/A	0.5	0
Clifford's pH 3.0	Reduce pH by conc. HCl in an open beaker	45.9	10.1	N/A	35.8	5 – 6
		46.4	9.9	N/A	36.5	5 – 6
		47.5	9.4	N/A	38.0	5 – 6
Clifford's pH 8.3	None	45.2	23.3	N/A	22.9	1 – 3
		45.4	23.1	N/A	22.9	1 – 3
		46.2	23.5	N/A	23.4	1 – 3
Edwards Method	Reduce pH by 0.05% H_2SO_4 after filtration	46.0	16.8	17.2	32.0	5 – 10
		46.9	16.1	18.0	32.2	5 – 10
		48.7	19.5	19.1	26.2	5 – 10
		48.3	19.5	19.1	25.8	5 – 10

N/A- Not analyzed

* Particulate As V is the difference between "Feed As V" and "soluble As V in 0.2 µm filtrate" i.e., Column 3 minus Column 4.

†Air contact time is the time of sample exposed to air before preservation or speciation or the actual speciation in the minicolumn. Iron precipitation in the Clifford method at pH 3.0 and 8.3 indicates that oxidation occurs during speciation in the minicolumn.

Overall, the EDTA-HAc method was observed to be the best method for arsenic speciation as it successfully overcame the effects of dissolved iron interference. Moreover, the chemicals involved, HAc and EDTA are less hazardous than H_2SO_4 and HCl, and hence meet our criteria for using only relatively safe chemicals. On the basis of these speciation tests, the EDTA-HAc Method was chosen for further method development to convert it into an accurate and robust field arsenic speciation method.

FURTHER DEVELOPMENT OF THE EDTA-HAc METHOD

Optimizing the Resin Volume

Initially, 7-mL of chloride-form anion exchange resin was used to perform the HAc-EDTA arsenic speciation method and only 50 mL of the challenge water, containing either As III or As V was passed through the resin. However, it is important to know the ultimate capacity of the resin minicolumn in order to ensure that there is sufficient excess capacity to accommodate spikes in the influent water quality. Therefore, the capacity of the anion exchange resin using the EDTA-HAc Method was determined by passing 3000 mL of feed water through the resin minicolumn. Figure 3.41 shows the As V breakthrough from a 7-mL chloride-form resin minicolumn. A short breakthrough was noted, with the maximum column capacity being equal to 50 mL or 7 BV (Figures 3.41 and 3.42).

It was expected, based on previous studies (Ghurye, Clifford and Tripp 1999), that the column throughput, based on the feed sulfate concentration of 50 mg/L, would be at least 800 BV (5600 mL) in the absence of HAc and EDTA. Two effects are probably at work here to shorten this predicted capacity: low pH and EDTA-metal-arsenate complex formation. Near pH 4.0, arsenate is removed as the monovalent ion, $H_2AsO_4^-$, which is far less preferred than the divalent ion, $HAsO_4^{2-}$, removed at neutral-and-higher pH (Horng and Clifford 1997). The second reason for the greatly attenuated capacity could be the formation of Fe-EDTA-As complexes in solution. Such complexes would not represent preferred species (compared with divalent anions such as arsenate (As V) or sulfate) by an SBA resin (Dudzinska and Clifford 1991).

Anions such as bicarbonate, which are less preferred that the presaturant ion (chloride for chloride-form resin), typically exhibit the convex-with-respect-to-X-axis shape of the elution curve seen in Figures 3.41 and 3.42. Therefore, to afford a margin of safety, resin volumes of 10 and 15 mL were tested by passing EDTA-HAc-stabilized As V-containing challenge water. The results are shown in Figures 3.43 and 3.44. The respective throughputs before As V breakthrough were 130 mL for 10-mL resin volume and greater than 150 mL for 15-mL resin volume. Since, the EDTA-HAc method called for a throughput of 60 mL, the resin volume was fixed at 10 mL, which would provide two-and-a-half times the target 60-mL capacity.

In order to investigate the reason for the short (7 BV) As V run length of ground waters containing EDTA, Fe II, and HAc, three minicolumn runs were performed using different combinations of Fe, EDTA, and HAc in NSF Challenge Water with 50 ppb As V. The results are plotted in Figure 3.45.

Operating at pH 8.3 with iron and EDTA (no HAc), a very long run length of more than 200 BV was achieved, and apparent iron uptake by the resin was visually observed as an expanding reddish brown zone near the column inlet. The red-brown zone near the column inlet at pH 8.3 can be explained in light of Fe-EDTA complexation chemistry. EDTA forms 1:1 water-soluble complexes with Fe II and Fe III. The stability of the complexes depends on the form of iron and the pH of the solution. The Fe III-EDTA complex becomes hydrated and brown colored at pH >6.5 (Johnson 1958). At pH 10.5, the hydrated Fe III-EDTA complex is disrupted and forms hydrated ferric oxides precipitates. Although the experiment was started with Fe II, oxidation to Fe III was expected, because Fe II is rapidly converted to Fe III at pH 8.3, especially upon exposure to the atmosphere. So, at pH 8.3 with EDTA and Fe III, the reddish-brown resin observed near the column inlet was probably due to the uptake the brown hydrated anionic EDTA complex $[Fe(OH)_2EDTA]^{3-}$.

When the pH of the feed was reduced from 8.3 to 3.5 using HAc and in the absence of iron, the As V run length was shortened 10-fold to about 20 BV. See Figure 3.45. This drastic shortening of the run length is expected based on the predominant arsenic V species being $H_2AsO_4^-$, which has a very low affinity for the resin.

When the pH was reduced to 3.5 by adding HAc to the challenge water containing EDTA and iron, the reddish-brown iron zone previously seen at pH 8.3 was not observed on the resin, and the As V broke through almost immediately at about 7 BV. This suggests that at pH 3.5, a colorless complex comprising Fe II, EDTA, and $H_2AsO_4^-$ may be formed, and that it has less affinity for the resin than does the monovalent ion $H_2AsO_4^-$, which breaks through at about 20 BV under the same conditions in the absence of Fe II (Figure 3.45).

To circumvent EDTA and achieve longer run lengths, 0.01 M HAc alone was examined to complex iron and eliminate its interference in arsenic speciation. The resin did not show any capacity for As V (Figure 3.46) because HAc was not as good a ligand as EDTA, and therefore, was not able to complex iron and avoid its interference. This validated the use of EDTA to complex iron.

Evaluating Other SBA Resins

Four different SBA resin types were evaluated for the EDTA-HAc method. These were:

1. SBA Resin 1[9]
2. SBA Resin 2[10]
3. SBA Resin 3[11]
4. SBA Resin 4[12]

The purpose for evaluating standard-mesh commercially available resins was to greatly lower the cost of the resin. Resin 1 is the expensive ($350/kg) fine mesh resin that has been used throughout the study, and Resins 2-4 are inexpensive ($20/kg) commercially available resins in standard 16×50 (1.18×0.300 mm) mesh size. The comparison of these four resins is shown in Figure 3.47. Of the four resins studied, resins 1 and 2 exhibited maximum capacity. Three resins gave acceptable As V retention and could be used in the minicolumn to separate As III from As V by the EDTA-HAc method. To verify that the ASB-2 and A850 resins could be used for speciation, additional tests would have to be run to verify >90% passage of As III.

---

9. Dowex IX8 50-100 Mesh, Supelco, Bellefonte, PA
10. Ionac ASB2 16-50 Mesh
11. Purolite A850 16-50 Mesh
12. Rohm and Haas A402 16-50 Mesh

Table 3.7
Effect of flow rate on As V retention on resin*

Operator	Flow rate (ml/min)	Feed As V (µg/L)	Effluent As V (µg/L)†	% As V retention
A	20	51.4	0.2	99.6
A	20	51.2	0.0	100.0
B	150	52.7	2.6	95.0
B	150	51.1	0.0	100.0

* 50×100 mesh Chloride-form Dowex IX8 SBA Resin
† 45-mL composite sample was collected after wasting initial 15 mL

Optimizing EDTA-HAc Concentrations

Initially 0.0005 M EDTA and 0.01 M HAc were used to complex iron and reduce the pH to eliminate problems due to iron precipitation. However, to ensure iron complexation in waters that may contain even higher concentrations of reduced iron or unusually high hardness levels (Ca^{2+} and Mg^{2+}), the EDTA and HAc concentrations were increased to 0.0172 M (500 mg/L) as the disodium salt and 0.087 M, respectively. The higher levels of EDTA and HAc have been used by researchers at the EPA for *preservation* of inorganic arsenic species (Gallagher et al. 2001). The maximum throughputs achievable for both combinations of EDTA-HAc are shown in Figure 3.48. The higher concentrations of added reagents provided an As V capacity of 85 mL (30 mL less than at the lower EDTA-HAc concentration), which is still 67% more than the 45 mL sample to be collected (after wasting the initial 15 mL effluent).

Flow Rate Experiments

In order to test the robustness of the EDTA-HAc speciation method using chloride-form resin (hereafter referred to as the Clifford EDTA-HAc Method), arsenic speciations were performed using two operators, referred to as Operator A[13] and Operator B[14]. Operator A performed the speciation at a typical flow rate of 20 mL/min, whereas Operator B manually forced the EDTA-HAc stabilized challenge water at a rate of 150 mL/min, as fast as physically possible, through the resin minicolumn. There wasn't any significant difference in either the As III passage or As V retention on the resin at these two significantly different flow rates. The results are shown in Table 3.7.

Effect of DMA and MMA

Experiments were performed to determine the fate of organic compounds that might be present in some surface waters. Dimethylarsinic Acid (DMA) and Monomethylarsonic acid

---

13. Saqib Karori, University of Houston Graduate Student
14. Ganesh Ghurye, University of Houston Graduate Student

Table 3.8
ICP-MS results for arsenic-spiked challenge water

As species	Arsenic spikings (µg/L)	Arsenic analyzed (µg/L)	% recovery
As V	50	46.8	93.6
	50	46.1	92.2
As III	50	44.2	88.4
	50	44.5	89
MMA	5	4.9	98
	5	4.9	98
DMA	5	5.5	110
	5	5.4	108

(MMA) were used as representative organoarsenicals in this study. In separate experiments, it was determined that MMA could be analyzed using the usual HGAAS method but not DMA (see Appendix F). Therefore, samples containing MMA or DMA were analyzed using ICP-MS. Both As III and As V samples were spiked with 20 µg/L of DMA or MMA, and the resulting solutions were stabilized with EDTA-HAc, and speciated as usual.

Before analyzing the speciated samples on ICP-MS, As III, As V, MMA, and DMA were spiked in the NSF challenge water and the spiked samples were run on ICP-MS. The analyzed spike concentrations are shown in Table 3.8. The results of analysis for the speciated samples to determine the fate of organoarsenicals during inorganic arsenic speciation using ICP-MS are shown in Table 3.9.

The results of the organoarsenicals experiments were inconclusive. The pK_1 for MMA is 3.4, which is slightly lower that the EDTA-HAc-stabilized sample pH of 3.5. Therefore, the distribution between uncharged and charged MMA should be approximately equal. Based on the distribution of the charged species, it was expected that approximately 50% of the 20 µg/L MMA would be retained on the resin and the remaining 50% would be recovered along with As III. But MMA quantitatively passed through the resin minicolumn, suggesting that the resin-phase pH was lower than 3.4. When present with As V, DMA was not retained on the resin and was quantitatively recovered in the resin minicolumn effluent. This was expected based on the pKa (6.27) of dimethylarsinic acid, which would be a neutral species at the separation of pH of 3.5. However, when present with As III, DMA behaved exactly the opposite, and was quantitatively retained on the resin while As III was completely recovered in the resin minicolumn effluent. We have no explanation for this unexpected result other than analytical errors. Should the behavior of MMA and DMA become a significant issue in field speciations, these experiments would have to be redone.

Table 3.9
Fate of organoarsenicals during inorganic arsenic speciation

As species	Feed spikes (µg/L)				Feed analyzed As (µg/L)	Effluent analyzed As (µg/L)	Comment
	As III	As V	DMA	MMA			
As V + DMA	—	50	20	—	70.5	25.2	125% DMA passage
	—	50	20	—	72.0	25.9	
As V + MMA	—	50	—	20	71.7	17.4	88% MMA passage
	—	50	—	20	72.0	17.5	
As III + DMA	50	—	20	—	70.8	50.7	100% DMA retention
	50	—	20	—	70.5	51.9	
As III + MMA	50	—	—	20	62.6	65.9	100% MMA passage
	50	—	—	20	62.8	66.0	

All samples analyzed using ICP-MS

Table 3.10
Effect of sulfate concentration on As V retention (duplicate runs)

Sample no.	Sulfate concentration (mg/L)	Feed As V (µg/L)	Effluent As V (µg/L)	% As V retention
1	50	50.6	1.0	98.0
2	50	51.9	0.3	99.4
3	120	52.5	0.8	98.5
4	120	52.2	0.7	98.7
5	240	53.5	1.6	97.0
6	240	53.0	1.7	96.8

Effect of Sulfate Concentration

Experiments were conducted to evaluate the effect of sulfate concentration on resin capacity because sulfate is usually the background ion that limits arsenic removal run length. The sulfate concentration was varied from 50 mg/L to 240 mg/L in the NSF challenge water spiked with 50 μg/L As V. A 45-mL composite sample was collected after discarding the first 15 mL. Table 3.10 shows As V concentration in the effluent from the resin minicolumn.

From Table 3.10, it is clear that a near 5-fold increase in sulfate had a trivial effect on the arsenic capacity of the speciation column. In the previous experiments with sulfate levels of 50 mg/L (similar to sample numbers 1 and 2 in Table 3.10), arsenic breakthrough was reached within 100 mL of the speciated sample. With the increase of sulfate concentration to 240 mg/L, the arsenic run length was reduced, but still the total volume before breakthrough was higher than 60 mL, which was used for speciation. Possibly, divalent sulfate is not a preferred species at the high ionic strength of the sample resulting from EDTA and HAc addition.

DEVELOPMENT OF A FIELD SAMPLING PROTOCOL

Once the method was validated in the laboratory, a field sampling protocol was developed to speciate representative arsenic-contaminated waters in the USA. A schematic layout of the sampling kit is shown in Figure 3.49.

Each arsenic speciation sampling kit for the field verification of the EDTA-HAc method contained the following items:

1. One ion-exchange minicolumn with 10 mL chloride-form resin.
2. One 60-mL bottle marked "A" with 60 μL nitric acid.
3. One 250-mL bottle marked "C" with enough EDTA-HAc for a final concentration of 87 mM and 500 mg/L, respectively.
4. One 60-mL bottle marked "C-1."
5. One 60-mL syringe with Luer-lock tip.
6. One 6-inch long piece of ⅛-inch flexible tubing.

When possible, simultaneous analyses were performed at both University of Houston and cooperating laboratories (with Battelle analyzing the results from the Ohio speciation experiments only) and the results were compared. The procedure to prepare the field speciation kit and protocol to perform the subsequent field speciation test is detailed in Appendix G.

RESULTS OF FIELD TESTS

Four cooperating water utilities were contacted to perform the field-testing in order to validate the new EDTA-HAc speciation method. These utilities were located in Ohio, North Dakota, California, and New Mexico. Late in the project, the investigators became aware of an arsenic field-speciation project being conducted at the University of Minnesota (UM) on Minnesota well waters. A cooperative effort was established between the two universities in which UM doctoral student Mindy Erickson prepared the field speciation kits and speciated the field samples, which were analyzed at UH. In this way, 12 samples were collected from

Minnesota to evaluate the speciation efficiency in EDTA-HAc preserved samples. The field-speciated As III result based on separation by minicolumn ion exchange was compared with a laboratory-speciated As III result based on separation by differential-pH arsine generation (Lopez et al. 1992).

Field Results: Ohio

Six sampling well locations were visited by Saqib Karori (Graduate Student, University of Houston) and Lili Wang (Battelle). Funding for the Battelle portion of the field sampling was provided by USEPA (Tom Sorg, project officer). Saqib performed the speciation tests conforming to the new EDTA-HAc Method without particle filtration developed at Clifford's Laboratory at University of Houston, whereas Lili performed the modified Edwards-Battelle Method (referred to as Battelle Method) to speciate As III and As V after a the pre-filtration step. Table 3.11 compares the results of the HAc-EDTA Method and Battelle Method.

As seen from Table 3.11, UH (using the Clifford EDTA-HAc Method) and Battelle (using Edwards Method) produced comparable results for Total arsenic, As III and As V (by difference). Speciation results varied significantly for only one sample, the City of Heath Well 4. It is worth noting that the Heath Well 4 was the only well where significant particulate arsenic was detected.

Field-testing in the Columbus Ohio area produced arsenic analysis by two methods, which were eventually compared: (a) Field speciated using Clifford EDTA-HAc method, (b) Field Speciated using Battelle method. It should also be noted that the field speciated samples were analyzed by different methods, with UH using AA Hydride generation and Battelle using ICP-MS. UH and Battelle calculated As V concentrations as the difference between the total arsenic concentration and As III concentration.

Figures 3.50 and 3.51 compare UH speciation results with Battelle results for As total and As III, respectively. The 45-degree line signifies the line of equal relation. It can be seen that total arsenic is quantified similarly by both the methods, whereas there is little inconsistency, though not significant, between the As III concentrations.

Field Results: North Dakota, California and New Mexico

After completing the Columbus Ohio area sampling, EDTA-HAc speciation kits were sent to the following participating utilities with arsenic-contaminated drinking water wells:

a. City of Grandforks, ND (Triplicate speciations of two wells)
b. City of McFarland, CA (Triplicate speciations of two wells)
c. City of Albuquerque, NM (Triplicate speciations of two wells)

All samples were collected in 250-mL bottles, which contained enough EDTA and HAc to achieve the final concentrations of 0.0172 M (500 mg/L) and 87 mM, respectively and shipped to UH by overnight mail. Table 3.12 shows the results of the field speciation performed by UH. The field kits, along with the instruction handout (shown in Appendix G), were shipped from UH to the locations in early February 2002. Three triplicate speciation kits were provided for each well location. For unknown reasons, possibly misinterpretation of the instructions handout, we experienced problems with the first set of field speciation tests performed in California and New Mexico. For example, in the first set of samples shipped from California to UH via overnight mail,

Table 3.11
Comparison of EDTA-HAc and Battelle Method arsenic field speciation results

		1	2	(2-1)	3	4 = (1-3)	5	6	7	(7-6)	8 = (1-6)	9 = (5-7)	(9-8)		
Location	Sampling location	As (T) Bat. µg/L	As (T) UH µg/L	ΔAs (T) UH - As (T) Bat. µg/L	As (S) Bat. µg/L	As (P) Bat. µg/L	As (T) Stab. UH µg/L	As (III) Bat. µg/L	As (III) UH µg/L	Δ As (III) UH - As (III) Bat. µg/L	As (V) Bat. µg/L	As (V) UH µg/L	Δ As (V) UH - As (V) Bat. µg/L	Fe (total) mg/L	Fe (dissolved) mg/L
City of Heath Water Treatment Plant, Heath, OH	Well 4	14.5	11.5	-3.0	11.6	2.9	10.4	9.7	6.0	-3.7	1.9	4.4	2.5	0.5	0.1
	Well 5	29.9	32.0	2.1	33	ND	31.0	22.4	27.9	5.5	10.6	3.1	-7.5	1.3	1.3
	Well 6	35.7	34.2	-1.5	36.4	ND	34.3	28.9	32.3	3.4	7.5	1.9	-5.6	1.7	1.6
	Well 8	13.4	16.2	2.8	13.6	ND	15.6	10.3	14.5	4.2	3.3	1.1	-2.2	1.1	1.1
Licking Valley Middle School, Newark, OH	Influent to Treatment System	64.4	67.6	3.2	67.2	ND	66.1	61.7	59.6	-2.1	5.5	6.5	1.0	1.7	1.1
Licking Valley High School, Newark, OH	Influent to Treatment System	61.9	66.8	4.9	66.3	ND	66.1	63.3	63.7	0.4	3	2.4	-0.6	1.4	1.4

As (T) = Total Arsenic; As (S) = Soluble Arsenic; As (T) stab. = Total Stabilized Arsenic
UH – The samples were analyzed in University of Houston Lab using Hydride Generation FIAS Unit
Bat. – The samples were analyzed in Battelle using EPA Method 200.8 using ICP-MS for As analysis.
As V Bat. = As (Total) Bat. – As (III) Bat. {col. 8 = col. 1 – col. 6}
As V UH = As (Total) Stabilized UH – As (III) UH {col. 9 = col. 5 – col. 7}

Table 3.12
UH results of arsenic field speciation test performed at field sites in North Dakota, California, and New Mexico, using the new EDTA-HAc method

	Well location	As III µg/L	As V µg/L	Total As µg/L
ND	Aggasiz	14.0	2.4	16.4
		13.2	4.6	17.8
		13.9	3.4	17.3
	Strasburg	5.4	9.2	14.6
		5.6	9.8	15.4
		6.2	9.8	16.1
CA	Garzoli	1.0	9.7	10.7
		0.5	9.7	10.2
		0.6	9.9	10.5
NM	WWWLK02	24.5	18.1	42.7
		17.3	24.4	41.8
		17.9	24.4	42.3
	WWWPON06	13.8	34.7	48.5
		13.7	33.7	47.4
		13.4	34.3	47.7

resin beads were found in the speciated samples (Bottle C-1) indicating that sampling/speciation was improperly done. Thus, the first set of California sample results had to be thrown out. Results from Table 3.12 are from the second sampling. No problems were encountered with the speciation samples from North Dakota.

The triplicate field speciation test results in Table 3.12 are not compared with any other method and serve only to indicate that chemists, samplers, and operators can perform the EDTA-HAc Field Speciation Method, that the results are reproducible, and that inorganic As III and As V can be speciated in this way.

Table 3.13
Comparison of As III results for field-speciated vs. preserved lab-speciated samples from Minnesota drinking water wells

Sample ID	Time of 1st analysis in the lab after preservation in the field (h)	Conc. of As(III) in the field speciated sample (µg/L)	Conc. of As(III) in the preserved sample @ UH lab (µg/L)	Conc. of Total As (µg/L)	Conc. of As(V) (diff. of As(Tot)-As(III) (µg/L)
115406	49	5.21	5.27	5.31	0.10
673611	48	20.9	20.8	23.5	2.6
409252	49	51.1	51.2	55.0	3.9
196987	48	9.27	9.48	10.0	0.73
167758	46	24.5	24.5	26.5	2.0
675566	54	4.65	4.82	4.82	0.17
213999	51	72.9	72.7	90.2	17.3
476592	31	11.8	11.5	20.3	8.5
485717	30	14.1	14.9	15.6	1.5
444426	27	11.4	11.9	13.4	2.0
566900	28	3.63	3.74	3.77	0.14
151140	52	26.7	27.0	33.8	7.1

Field Results: Minnesota

The objective of the Minnesota tests was to compare field speciation of arsenic with lab speciation of the same EDTA-HAc preserved sample performed 1-2 days later. The same sampler, doctoral student Mindy Erickson, prepared all the field speciation kits, preserved all the samples and performed all the field speciation tests. This comparison was made easier by the fact that the same EDTA-HAc preserved sample could be immediately speciated by minicolumn ion exchange in the field and later speciated in the UH Lab using differential-pH hydride generation to determine As III and As Tot. Preserved samples were shipped from the Minnesota field sites to UH by overnight express mail, and were received in the laboratory within 54 h of collection and preservation in the field. Concentrations of As III in the field-speciated samples, As III and As Tot in the preserved samples, and As V calculated from the difference between As Tot and As III are tabulated in Table 3.13.

As III in the minicolumn effluent of the field speciated samples was determined by a total arsenic analysis using the conventional HG AAS procedure. As III in the preserved samples was determined by a HG AAS speciation procedure, which detects only As III at pH 5.0 in a citric acid/citrate buffer.

The graphical comparison of As III in the field-speciated samples vs. As III in the lab-speciated samples is shown in Figure 3.52, where it is seen that there is no significant difference between the samples: slope = 1.0044; intercept - 0.2314, and $R^2 = 0.9998$. A paired-t test at the 95% confidence level indicates that the values are not statistically different ($P < 0.0001$).

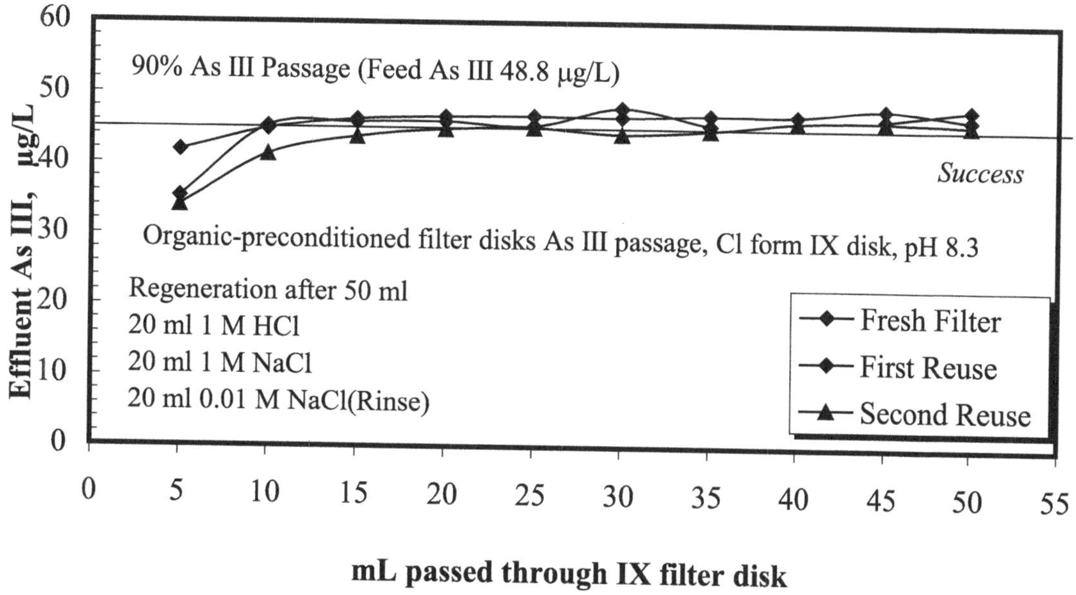

Figure 3.1 Arsenic III passage through chloride-form IX filter disk (fresh and regenerated) with organic preconditioning and HCl-NaCl pretreatment, feed pH = 8.3

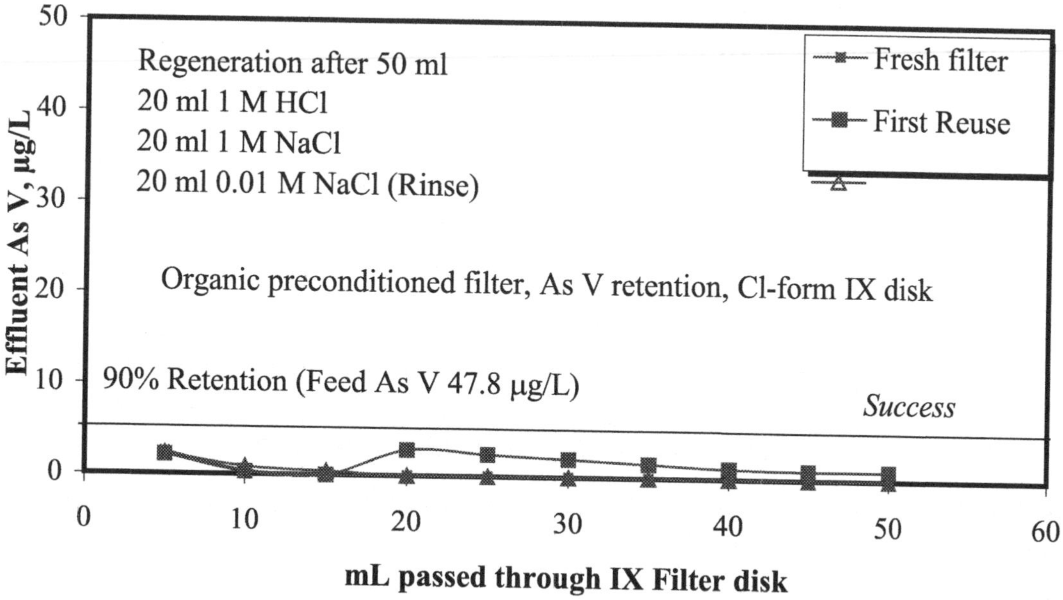

Figure 3.2 Arsenic V retention on chloride-form IX filter disk (fresh and regenerated) with organic preconditioning and HCl-NaCl pretreatment, feed pH = 8.3

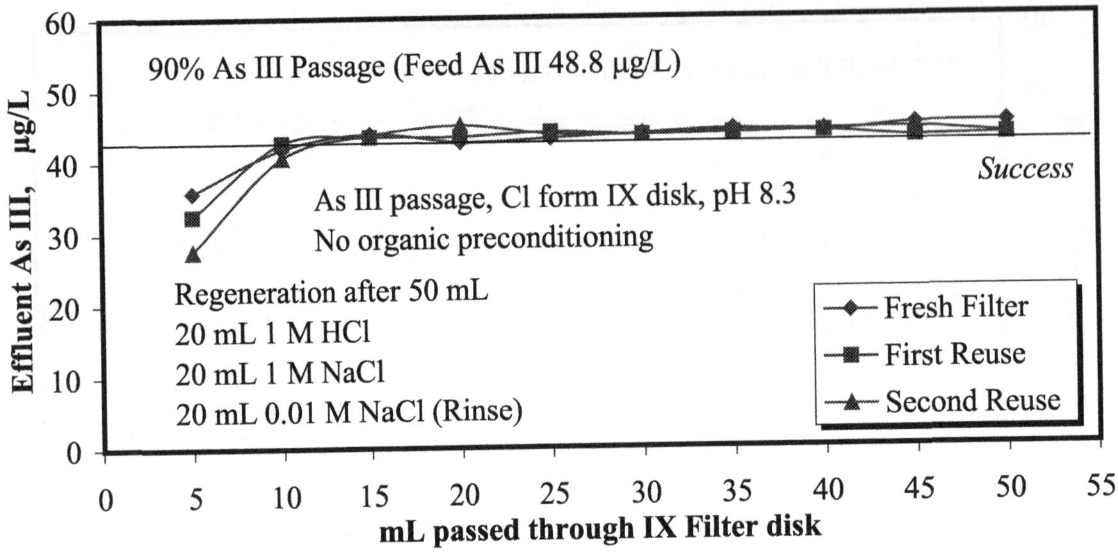

Figure 3.3 Arsenic III passage through HCl-NaCl pretreated chloride-form IX filter disk, no organic-solvent conditioning, feed pH 8.3

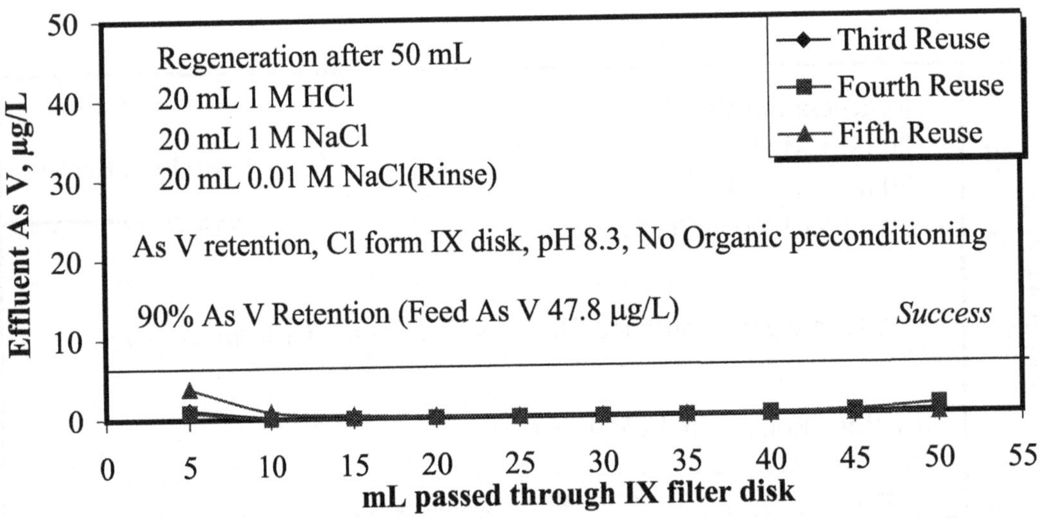

Figure 3.4 Arsenic V retention on HCl-NaCl pretreated chloride-form IX filter disk, no organic-solvent conditioning, feed pH 8.3

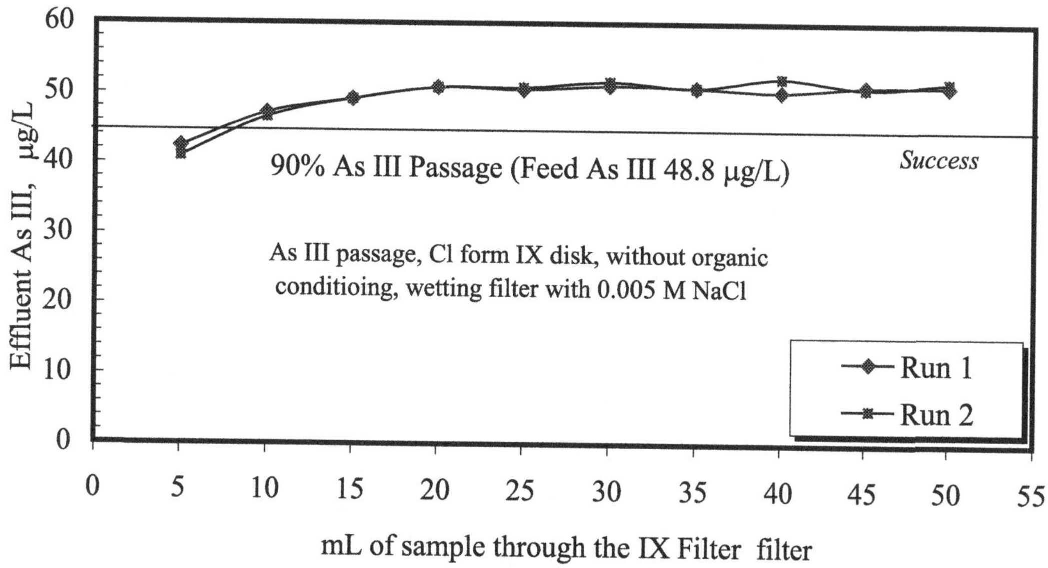

Figure 3.5 Arsenic III passage through a 0.005 M NaCl-wetted chloride-form IX filter disk, no organic-solvent or HCl-NaCl treatment, feed pH = 8.3

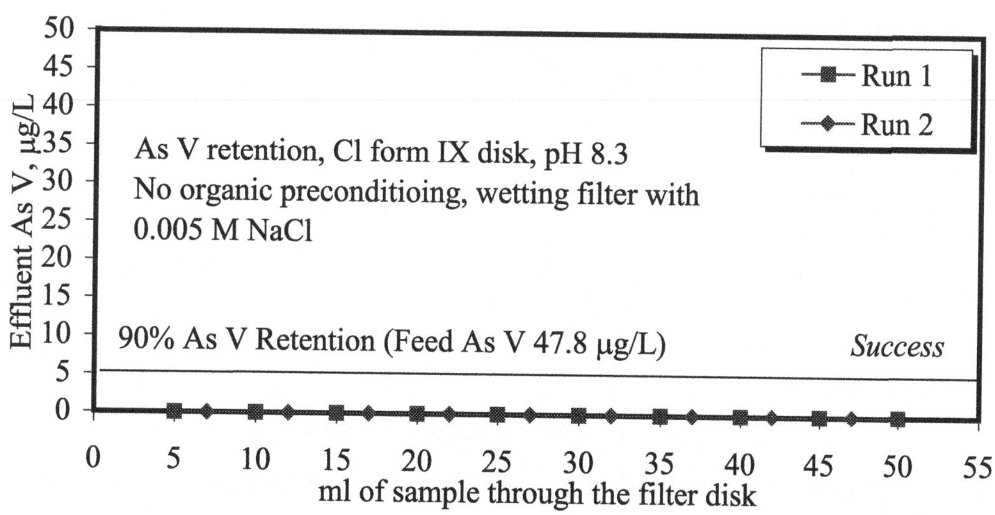

Figure 3.6 Arsenic V retention on chloride-form IX filter disk, no organic-solvent or HCl-NaCl treatment, feed pH = 8.3

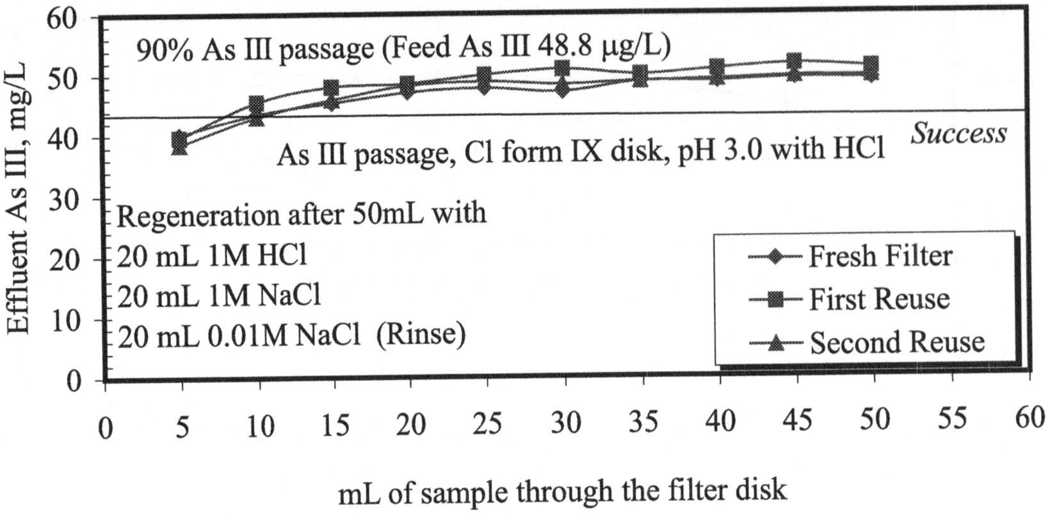

Figure 3.7 Arsenic III passage through chloride-form IX filter disk with subsequent regeneration and reuse, feed pH 3.0 using HCl

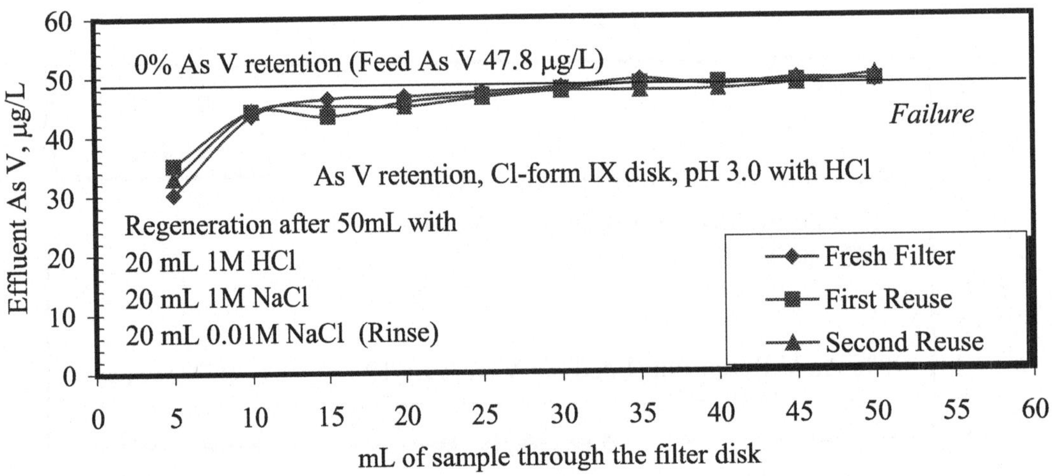

Figure 3.8 As V retention on chloride-form IX filter disk with subsequent regeneration and reuse, feed pH 3.0 using HCl

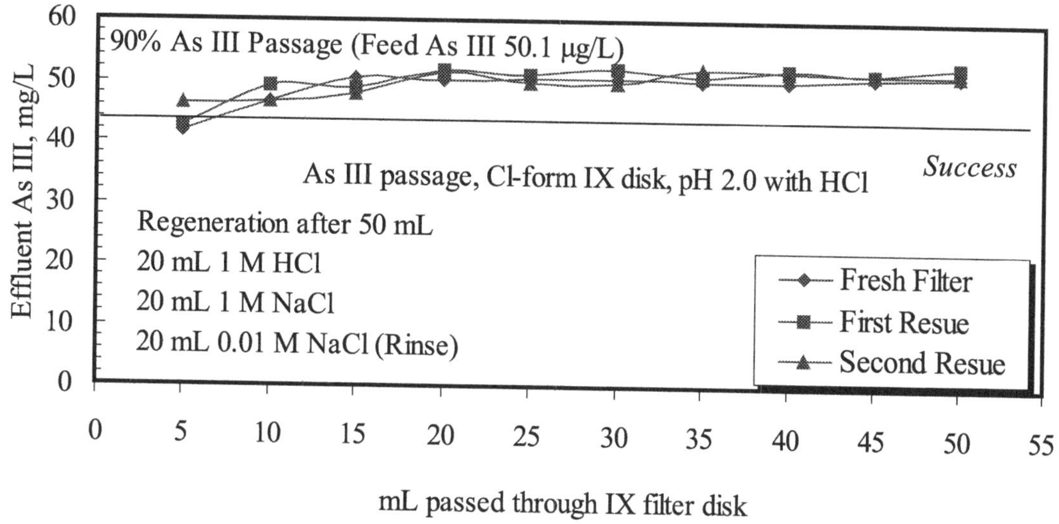

Figure 3.9 Arsenic III passage through chloride-form IX filter disk with subsequent regeneration and reuse, feed pH 2.0 using HCl

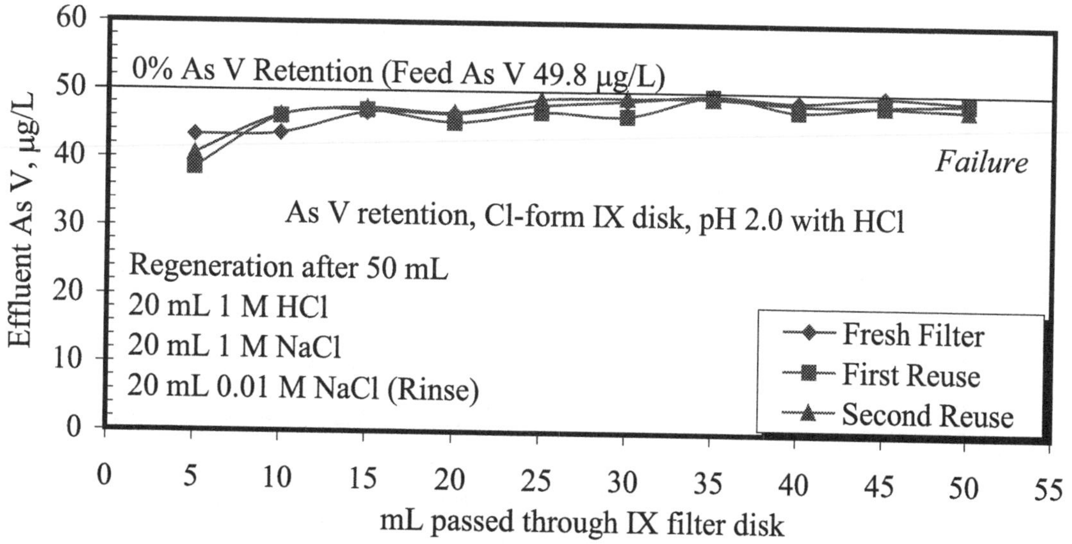

Figure 3.10 Arsenic V retention on chloride-form IX filter disk with subsequent regeneration and reuse, feed pH = 2.0 using HCl

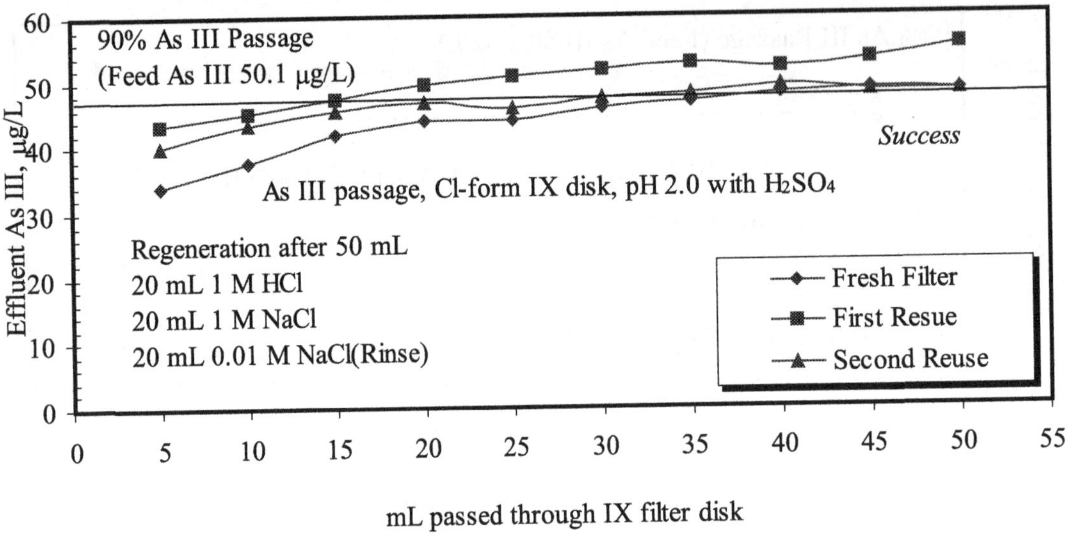

Figure 3.11 Arsenic III passage through chloride-form IX filter disk with subsequent regeneration and reuse, feed pH = 2.0 using H_2SO_4

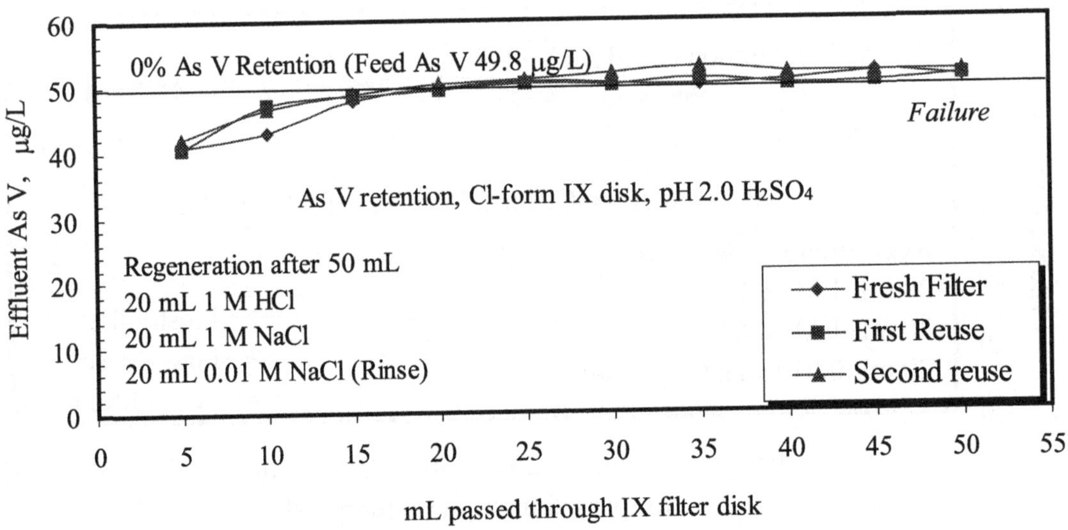

Figure 3.12 Arsenic V retention on chloride-form IX filter disk with subsequent regeneration and reuse, feed pH 2.0 using H_2SO_4

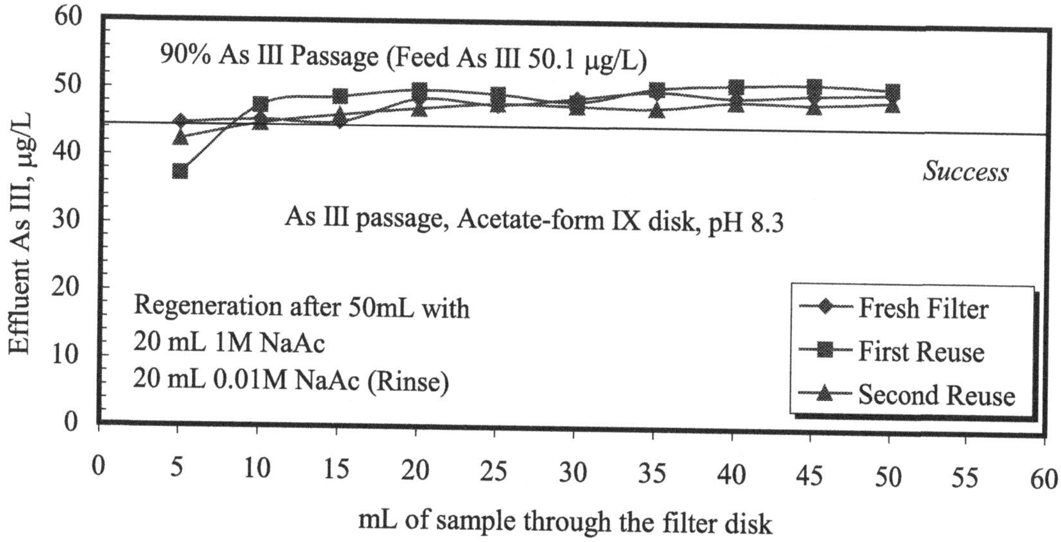

Figure 3.13 Arsenic III passage through acetate-form IX filter disk with subsequent regeneration and reuse, feed pH 8.3

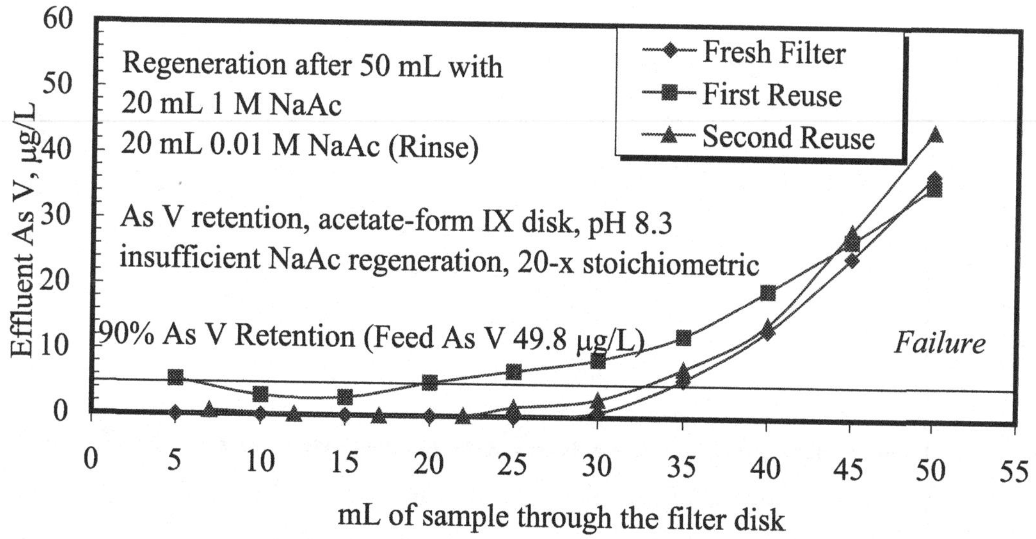

Figure 3.14 Arsenic V retention on acetate-form IX filter disk with subsequent regeneration and reuse, feed pH 8.3, effluent pH 4.2

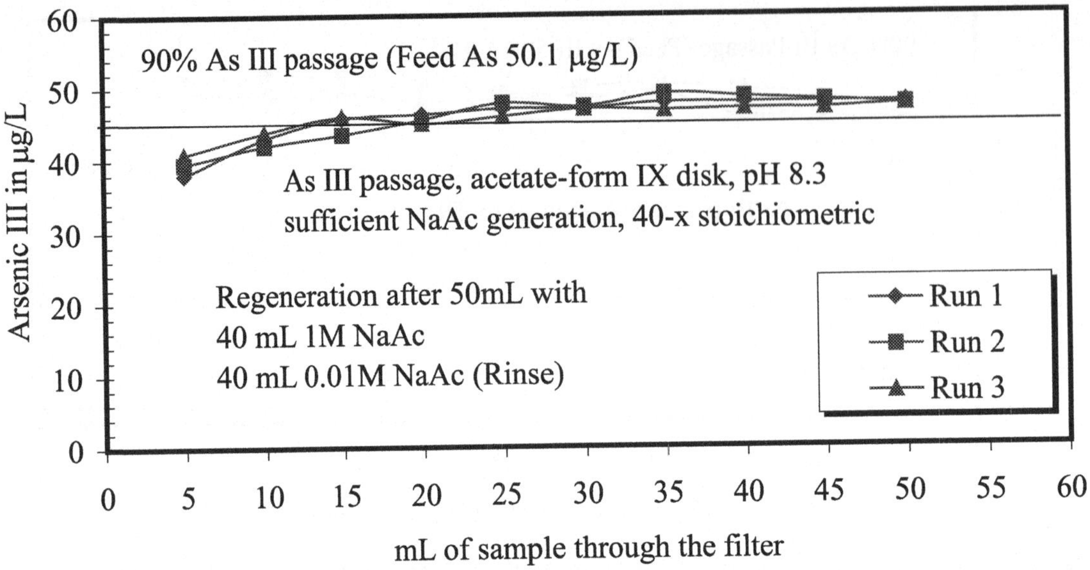

Figure 3.15 Arsenic III passage through acetate-form IX filter disk with subsequent regeneration and reuse, feed pH 8.3, effluent pH 4.2

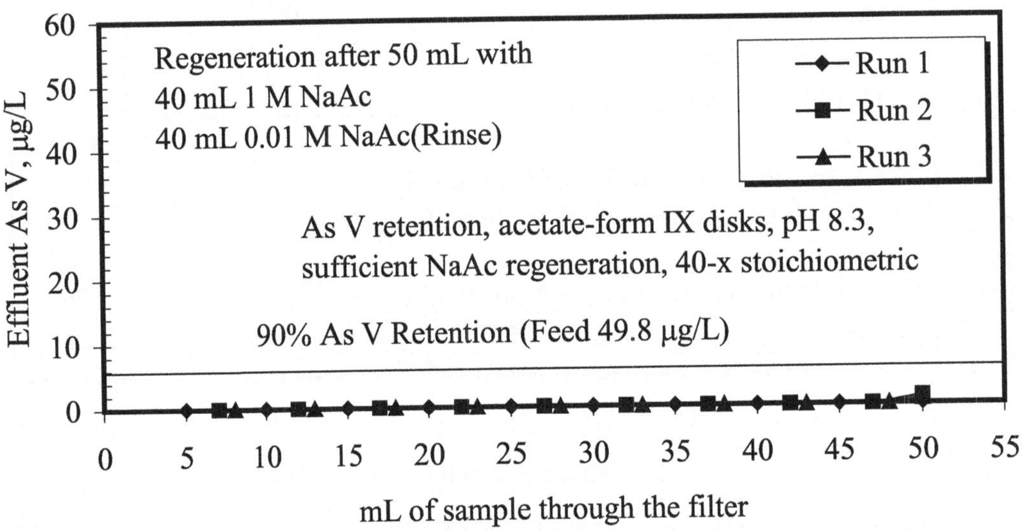

Figure 3.16 Arsenic V retention on acetate-form IX filter disk with subsequent regeneration and reuse, feed pH 8.3, effluent pH 4.2

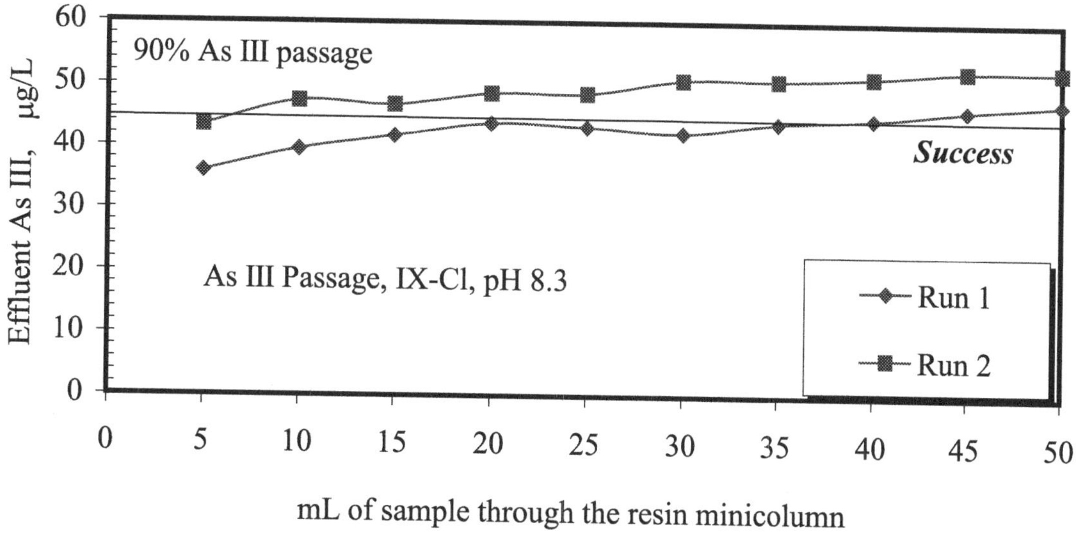

Figure 3.17 Arsenic III passage through chloride-form IX minicolumn, feed pH = 8.3

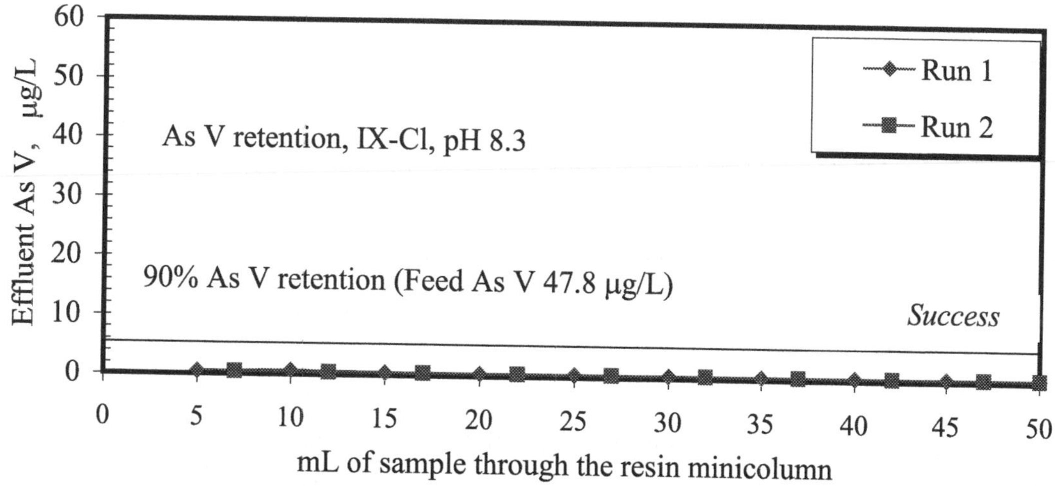

Figure 3.18 Arsenic V retention on chloride-form minicolumn, feed pH = 8.3

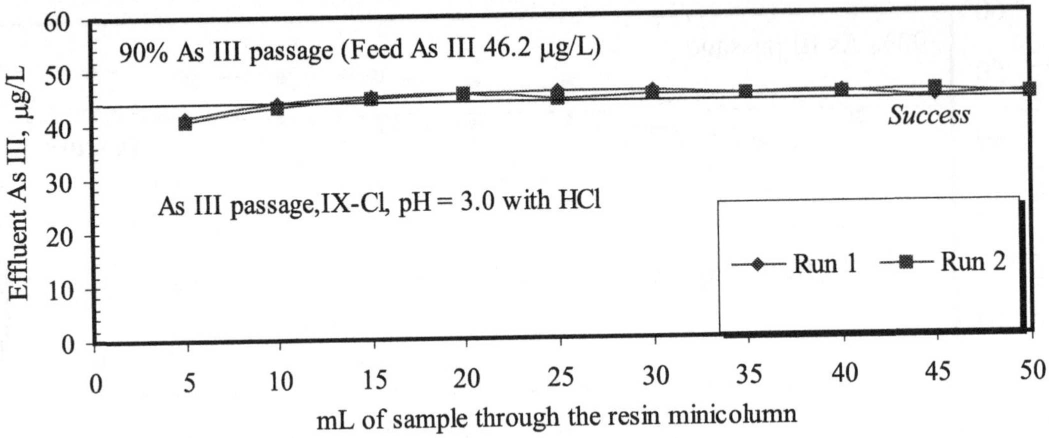

Figure 3.19 Arsenic III passage through chloride-form IX minicolumn, feed pH = 3.0 using HCl

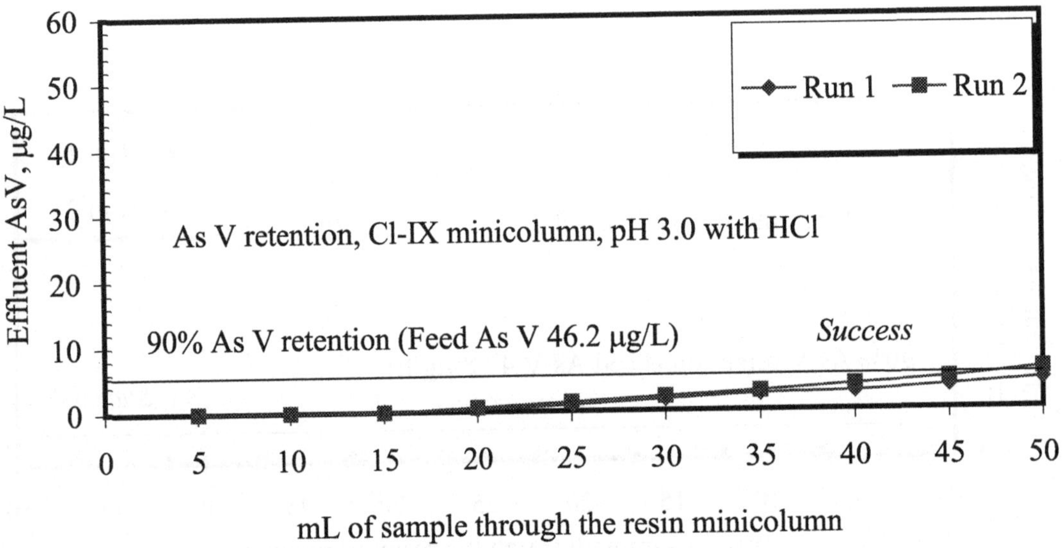

Figure 3.20 Arsenic V retention on chloride-form IX minicolumn, feed pH = 3.0 using HCl

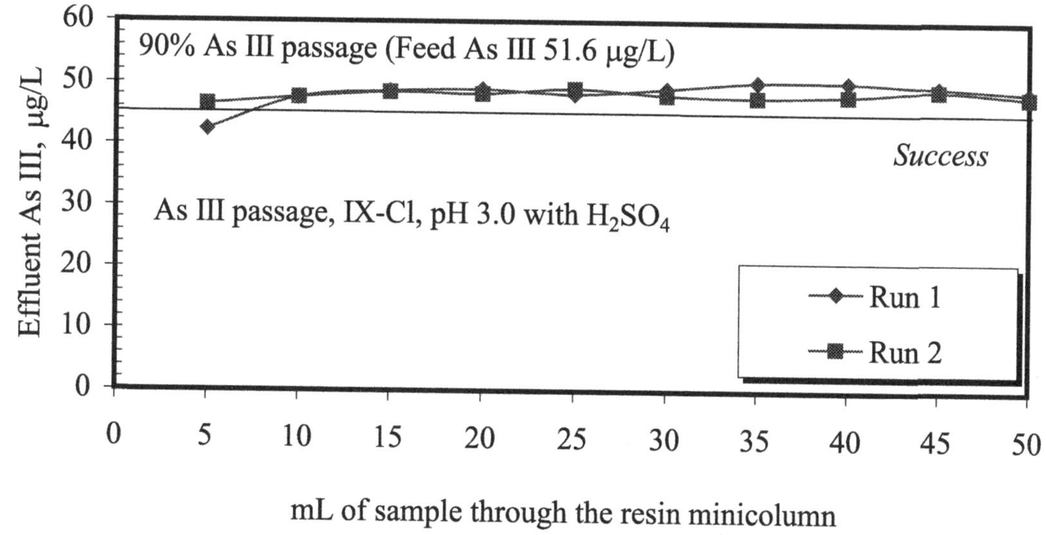

Figure 3.21 Arsenic III passage through chloride-form IX minicolumn, feed pH = 3.0 using sulfuric acid

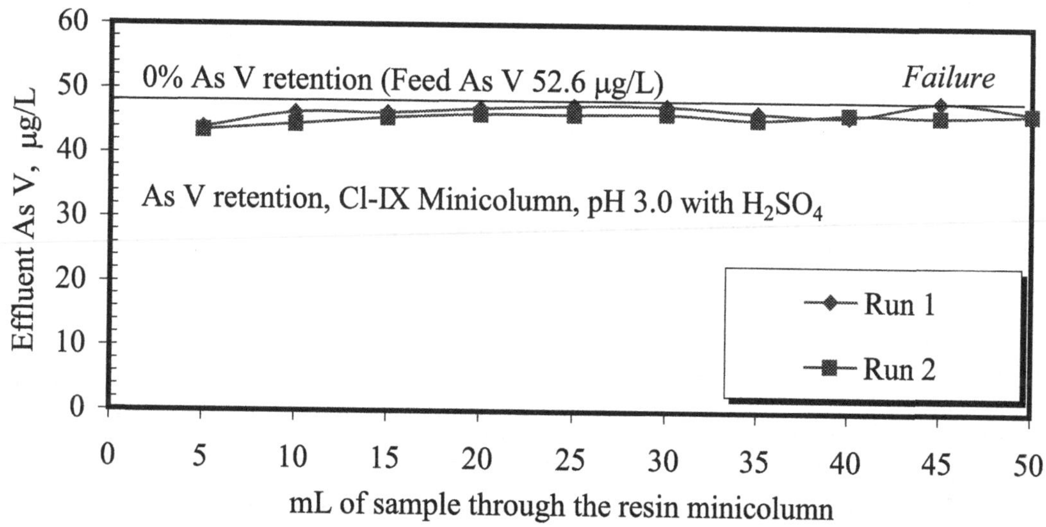

Figure 3.22 Arsenic V retention on chloride-form IX minicolumn, feed pH = 3.0 using sulfuric acid

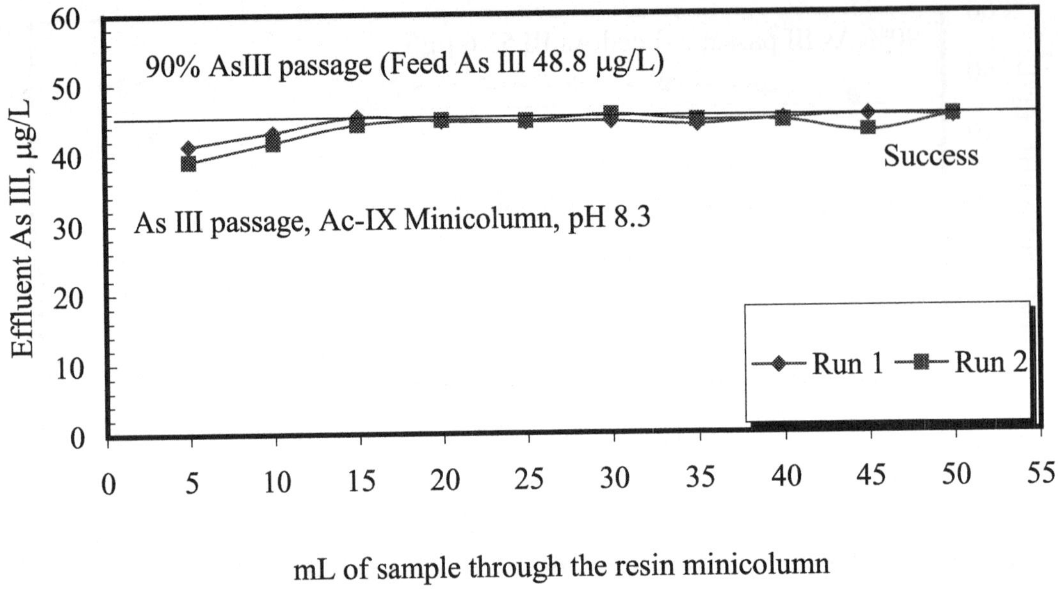

Figure 3.23 Arsenic III passage through acetate-form resin minicolumn, feed pH = 8.3, effluent pH 4.2

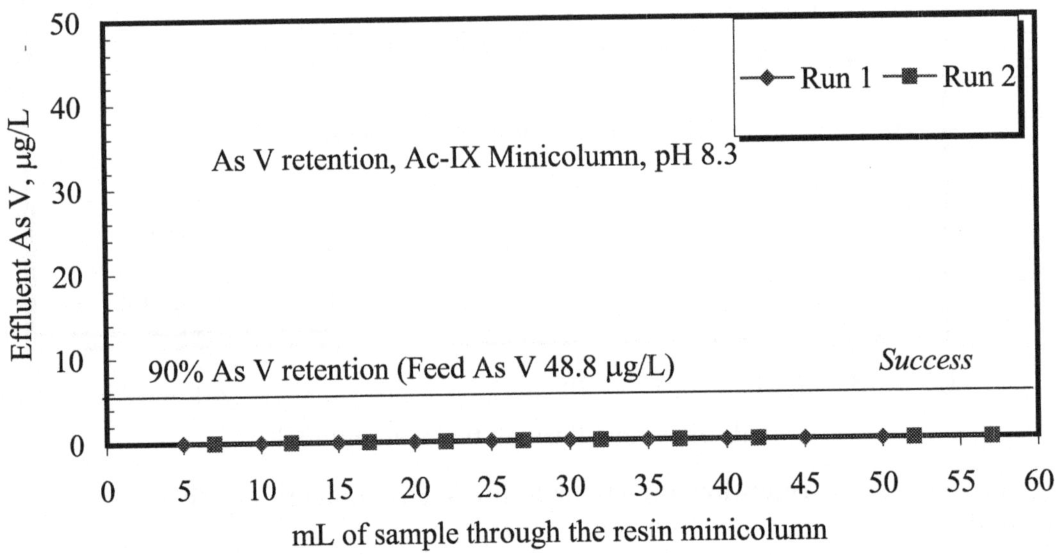

Figure 3.24 Arsenic V retention on acetate-form resin minicolumn, feed pH = 8.3, effluent pH 4.2

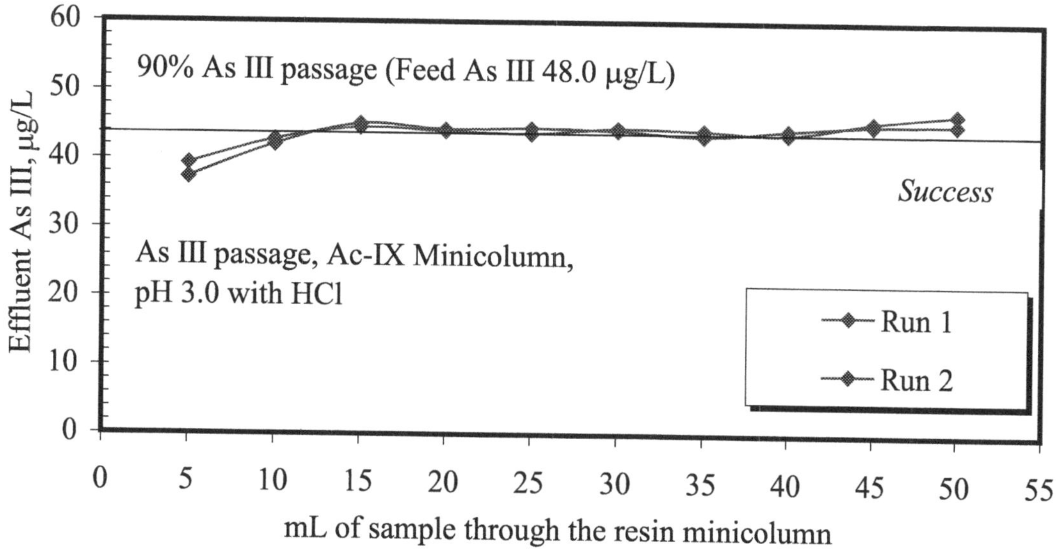

Figure 3.25 Arsenic III passage through acetate-form resin minicolumn, feed pH = 3.0 with HCl, effluent pH 3.8

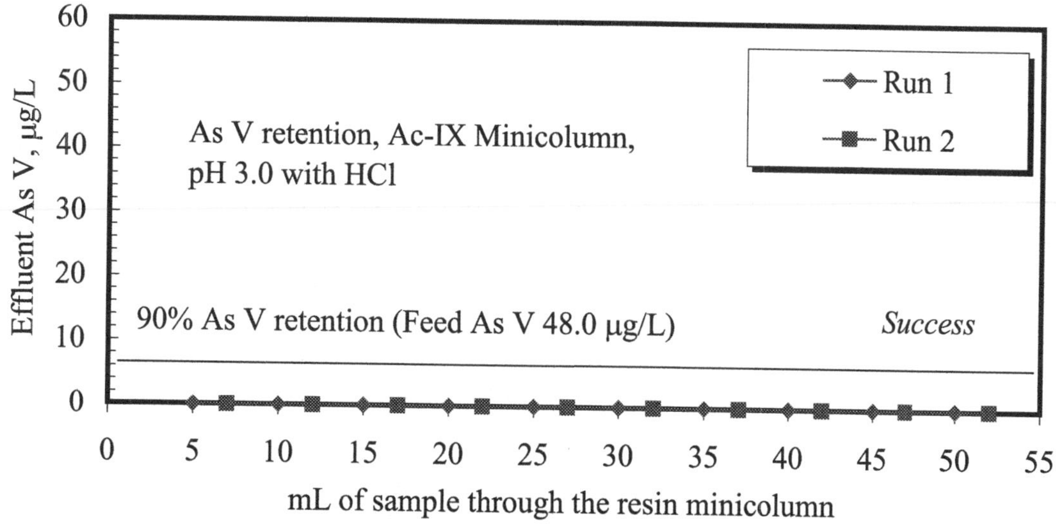

Figure 3.26 Arsenic V retention on acetate-form resin minicolumn, feed pH = 3.0 with HCl, effluent pH 3.8

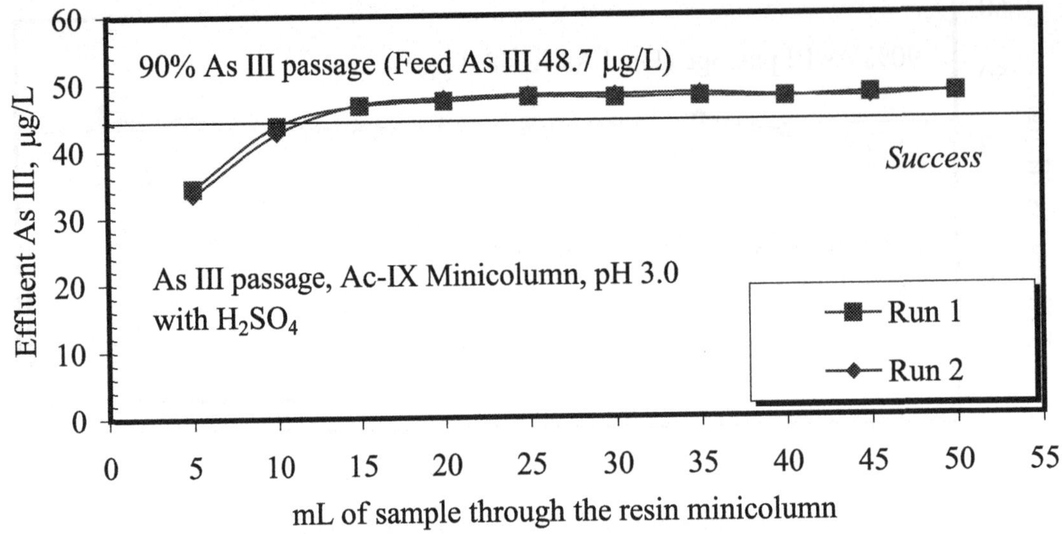

Figure 3.27 Arsenic III passage through acetate-form resin minicolumn, feed pH = 3.0 with H_2SO_4, effluent pH 3.9

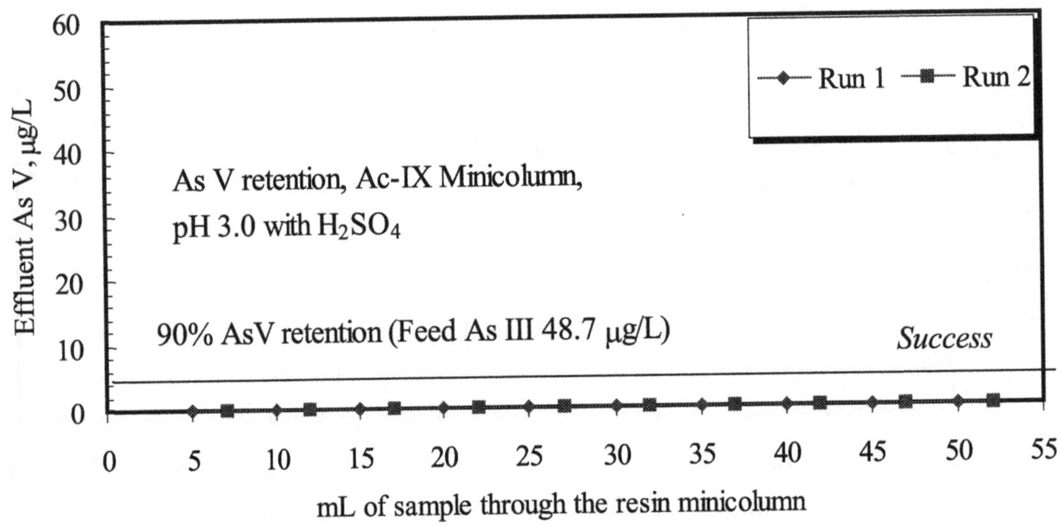

Figure 3.28 Arsenic V retention on acetate-form resin minicolumn, feed pH = 3.0 with H_2SO_4, effluent pH 3.9

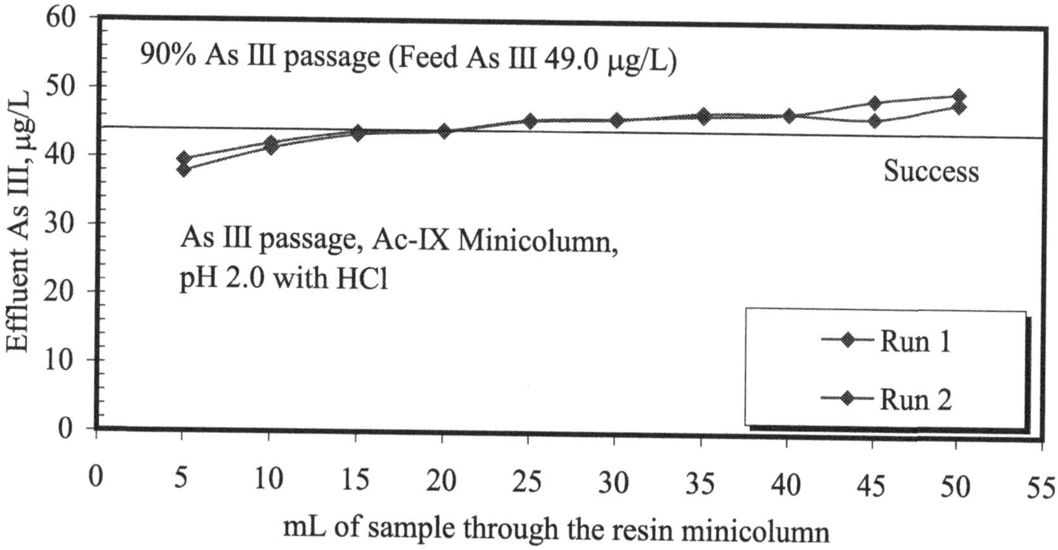

Figure 3.29 Arsenic III passage through acetate-form resin minicolumn, feed pH = 2.0 with HCl, effluent pH 3.8

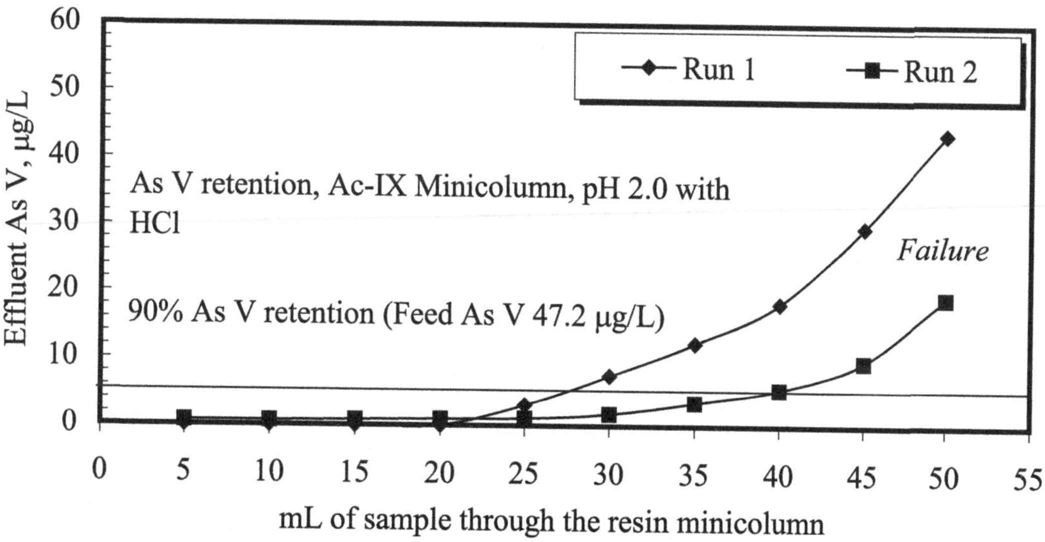

Figure 3.30 Arsenic V retention on acetate-form resin minicolumn, feed pH = 2.0 with HCl, effluent pH 3.8 (Resin volume was only 7 mL. More resin will give proportionally more As V capacity.)

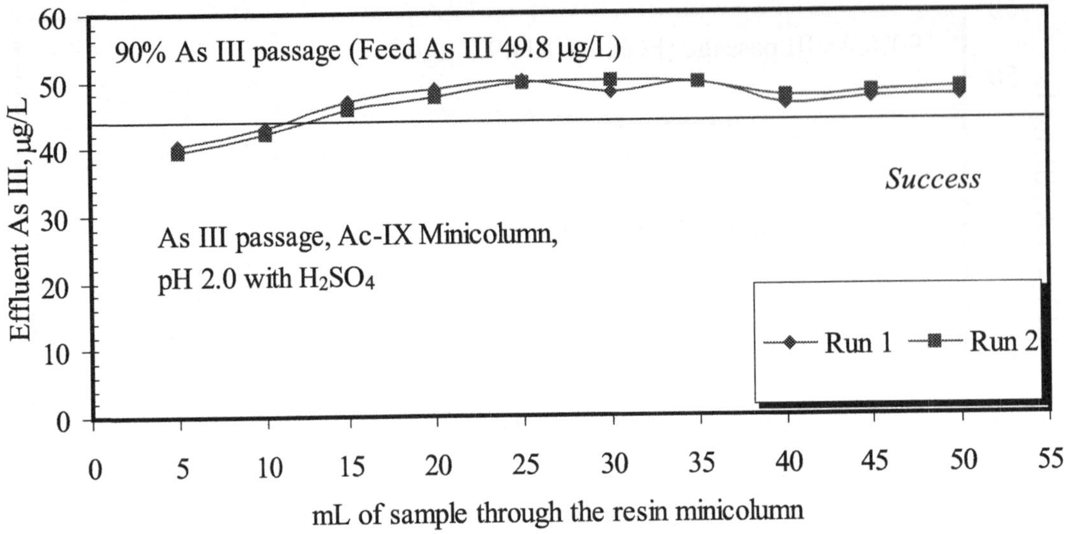

Figure 3.31 Arsenic III passage through acetate-form resin minicolumn, feed pH = 2.0 with H_2SO_4, effluent pH 3.8

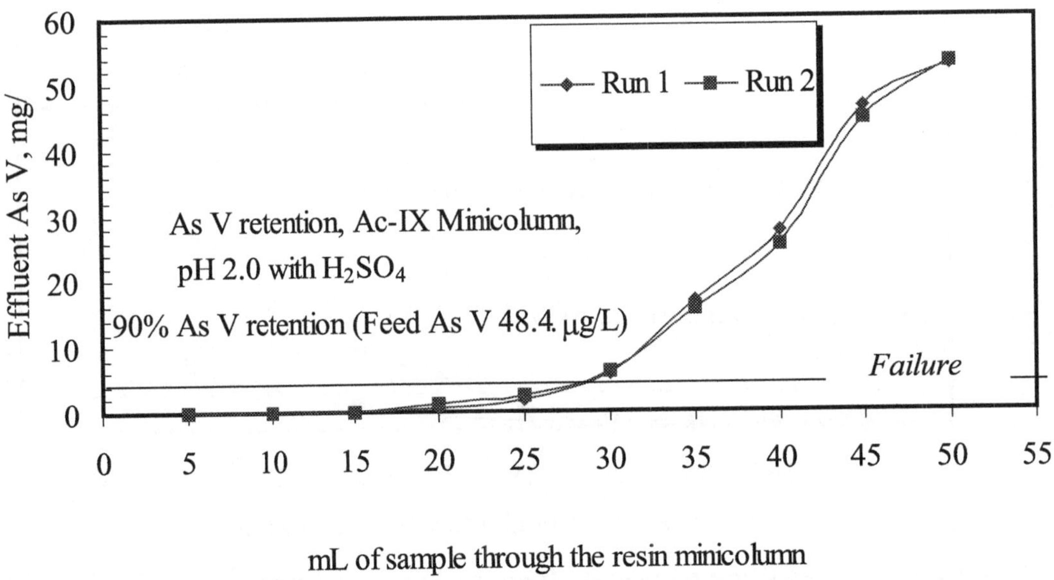

Figure 3.32 Arsenic V retention on acetate-form resin minicolumn, feed pH = 2.0 with H_2SO_4, effluent pH 3.8 (Resin volume was only 7 mL. More resin will give proportionally more As V capacity.)

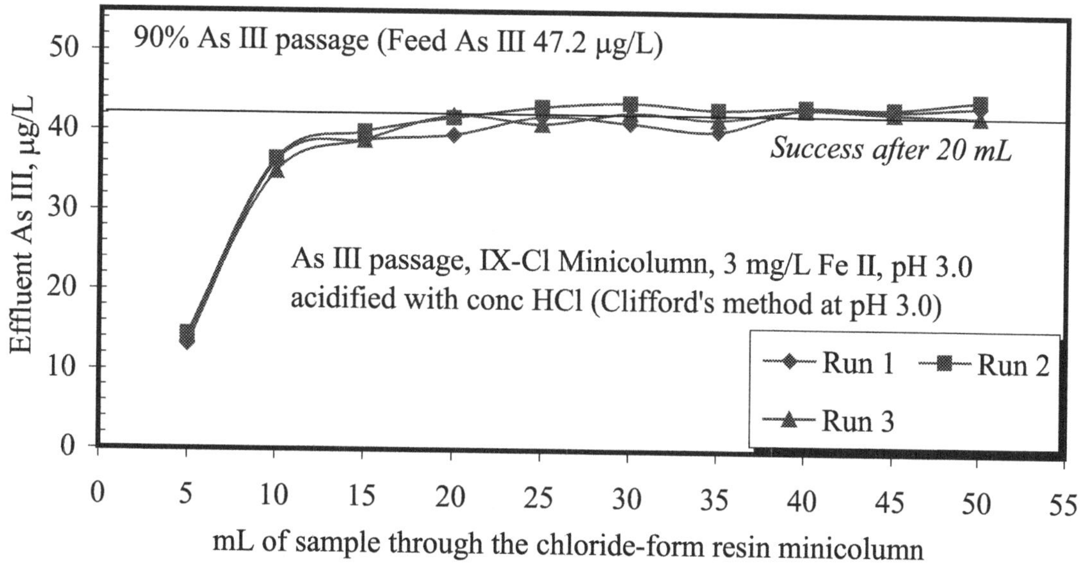

Figure 3.33 Arsenic III passage through chloride-form resin minicolumn with 3 mg/L Fe II at pH 3.0 acidified with HCl (Clifford's method at pH 3.0)

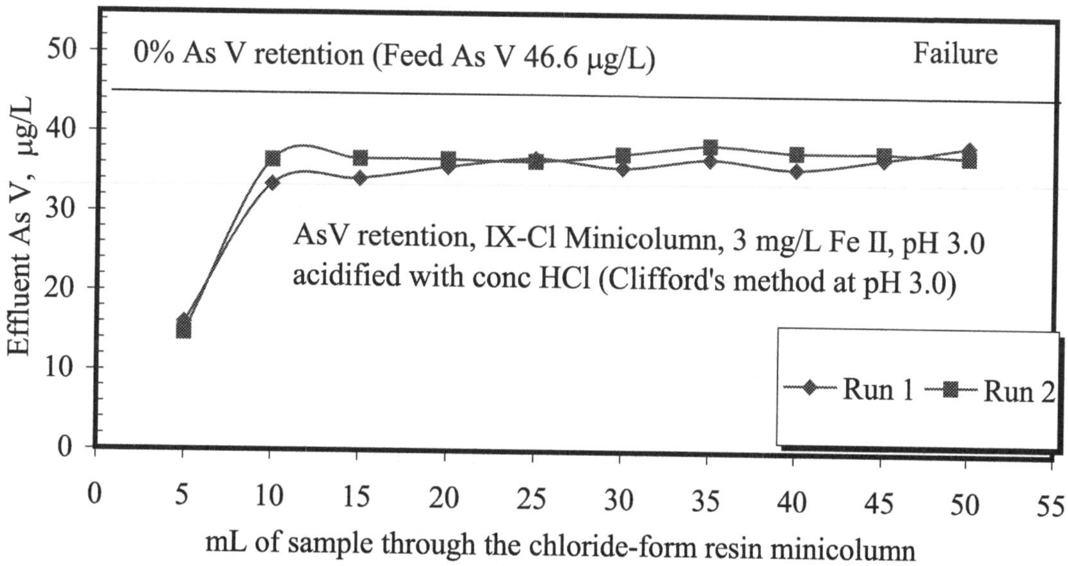

Figure 3.34 Arsenic V retention on chloride form resin minicolumn with 3 mg/L Fe II at a pH of 3.0 acidified with HCl (Clifford's method at pH 3.0)

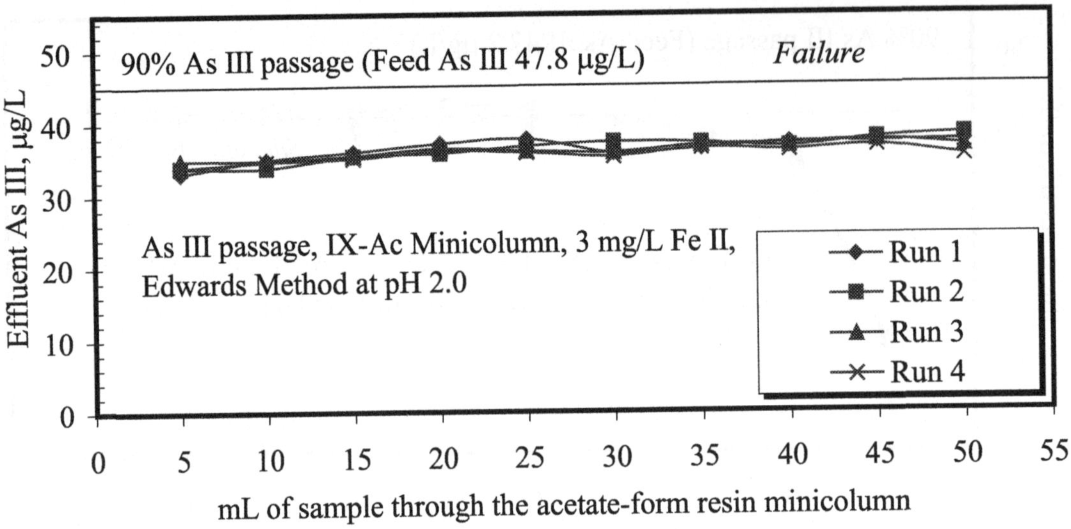

Figure 3.35 Arsenic III passage through acetate-form resin minicolumn with 3 mg/L Fe II at a pH of 2.0 using H_2SO_4 (Edwards Method)

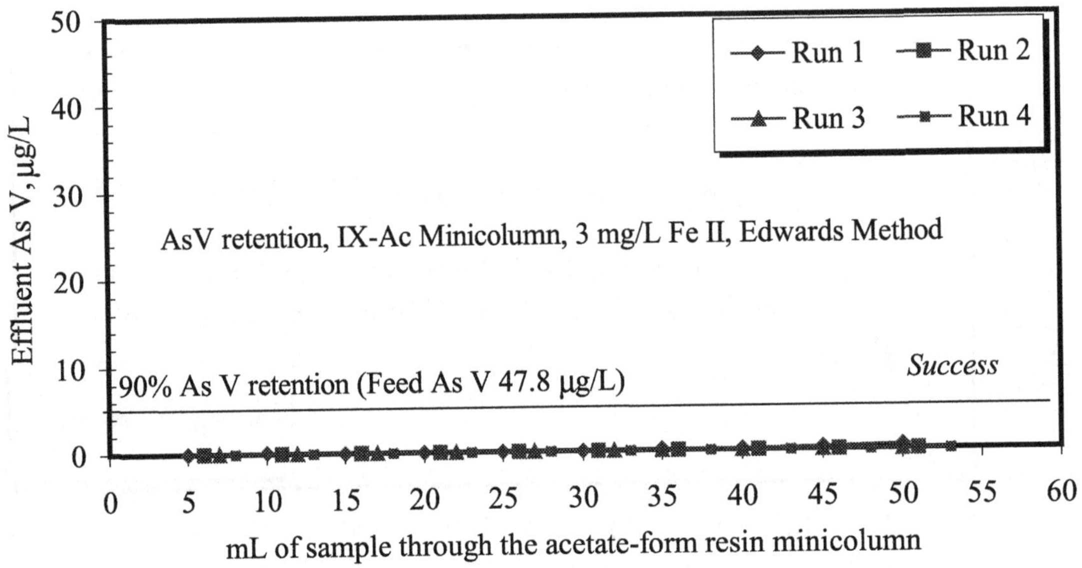

Figure 3.36 Arsenic V retention on acetate-form resin minicolumn with 3 mg/L Fe II at a pH of 2.0 using H_2SO_4 (Edwards Method)

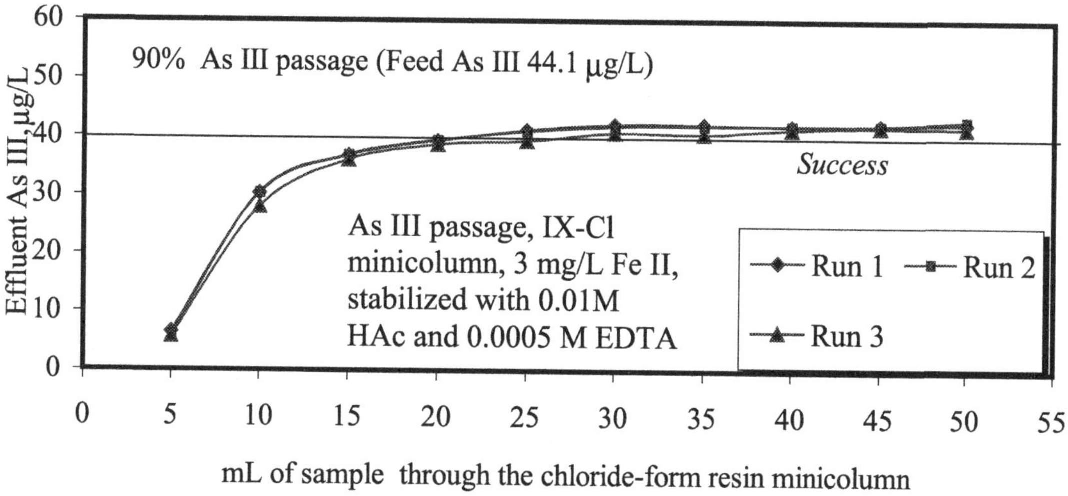

Figure 3.37 Arsenic III passage through chloride-form resin minicolumn with 3 mg/L Fe II and stabilized with 0.01 M HAc and 0.0005 M EDTA, feed pH 3.5, effluent pH 3.5

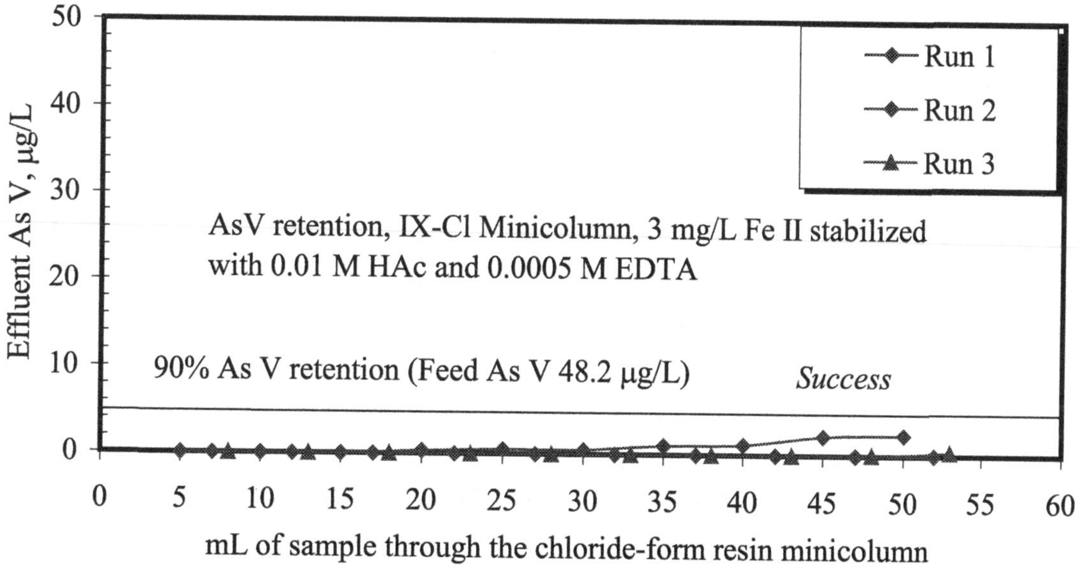

Figure 3.38 Arsenic V retention on chloride-form resin minicolumn with 3 mg/L Fe II and stabilized with 0.01 M HAc and 0.0005 M EDTA, feed pH 3.5, effluent pH 3

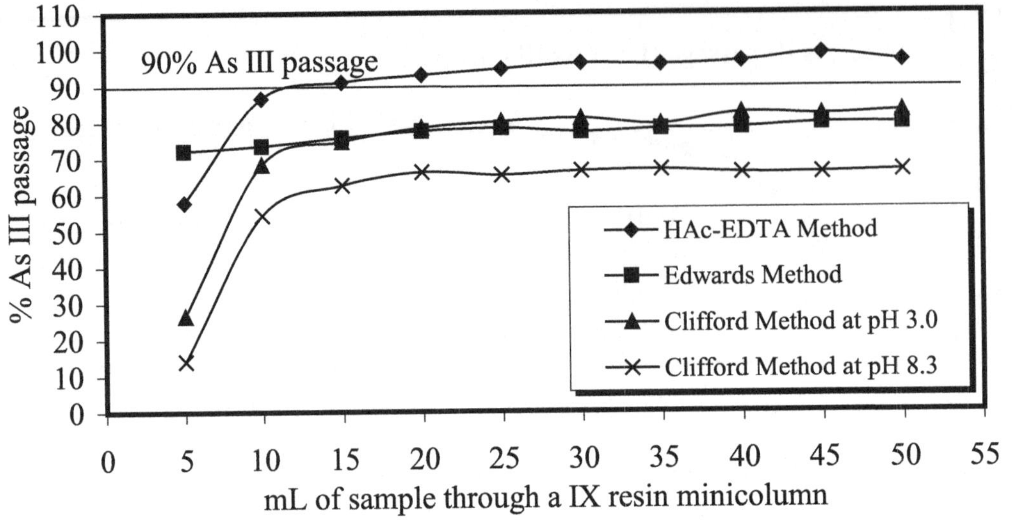

Figure 3.39 Comparison of percent As III recovery using four different methods in waters containing 2-3 mg/L Fe II

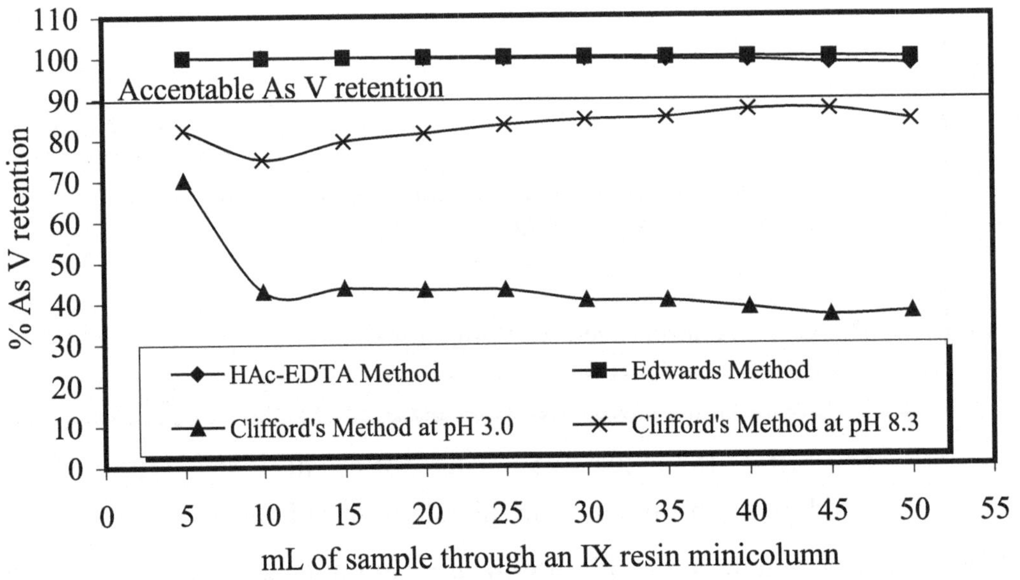

Figure 3.40 Comparison of percent As V retention using four different methods in waters containing 2-3 mg/L Fe II

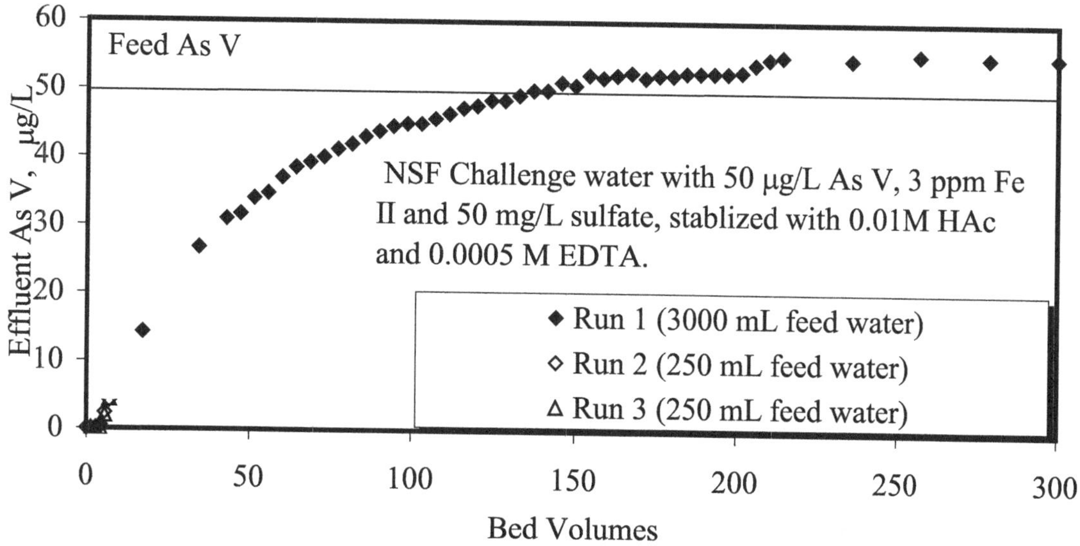

Figure 3.41 Complete breakthrough curve for 50 µg/L As V through 7-mL IX-Cl using 50-100 mesh Dowex resin

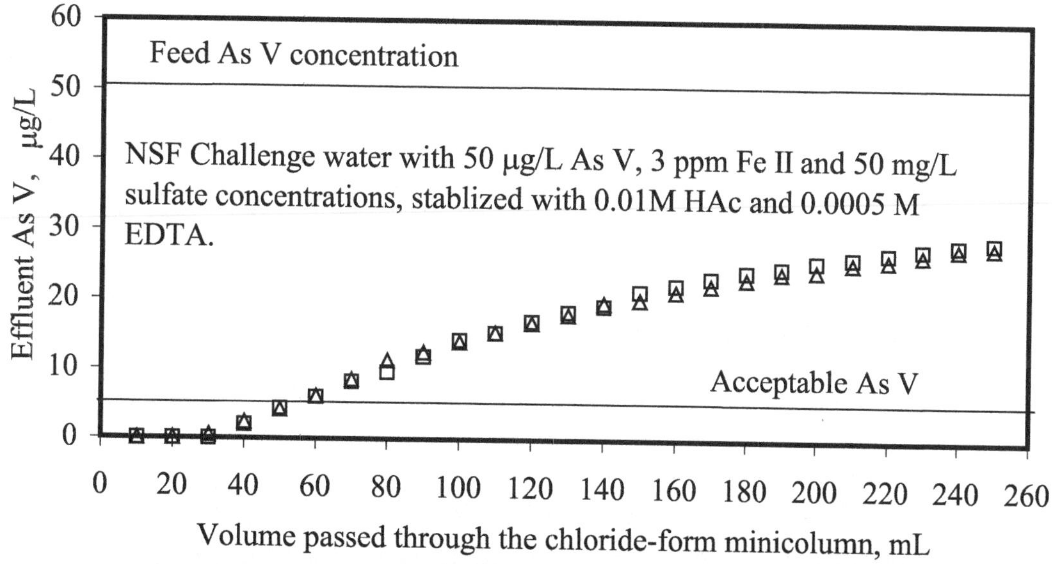

Figure 3.42 Early breakthrough curves for 50 µg/L As V through IX-Cl using 7-mL of 50-100 mesh Dowex resin (Note: 50 mL represents 7 bed volumes. This graph represents the first 34 BV (248 mL) of Figure 3.41 for run 2 and 3.)

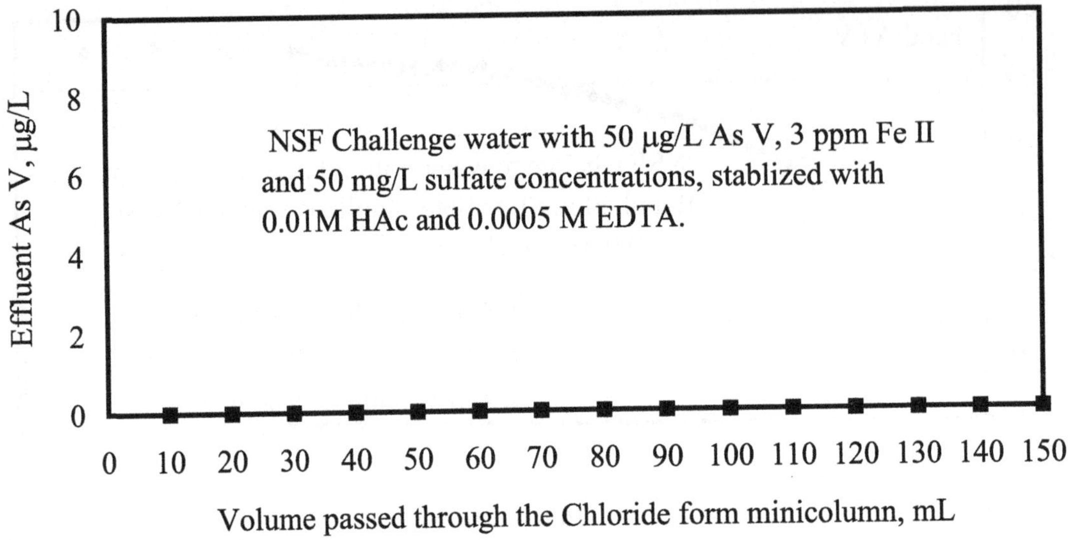

Figure 3.43 Breakthrough for 50 µg/L As V through 15-mL IX-Cl using 50-100 mesh Dowex resin

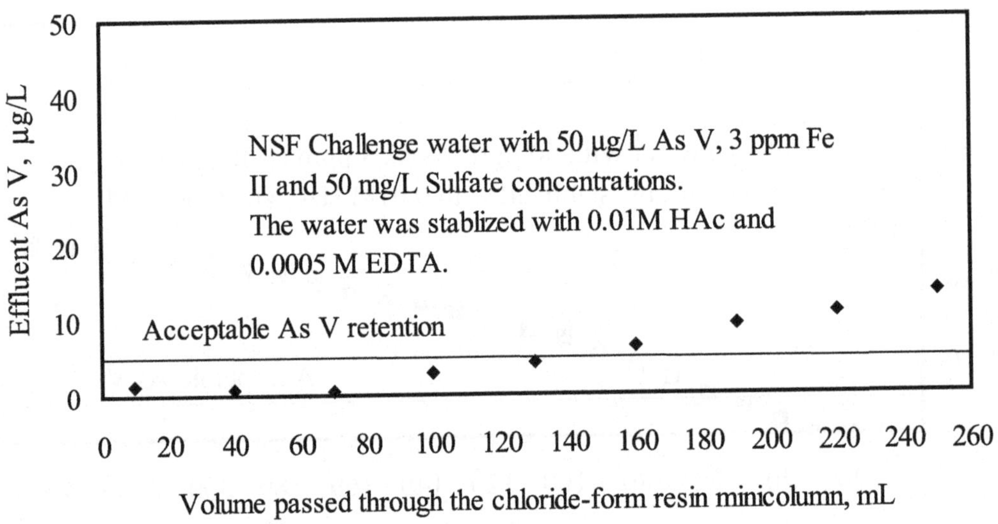

Figure 3.44 Breakthrough for 50 µg/L As V through 10-mL IX-Cl using 50-100 mesh Dowex resin

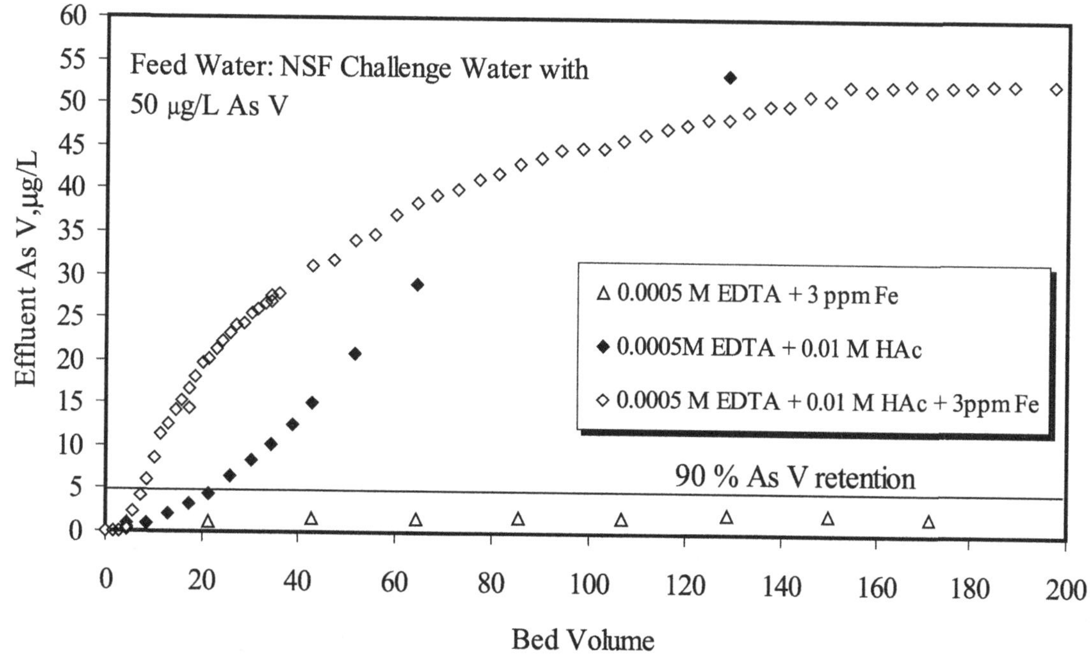

Figure 3.45 Breakthrough for 50 µg/L As V through IX-Cl using 7-mL of 50-100 Dowex resin with 3 mg/L Fe II at pH 8.3 and 3.5

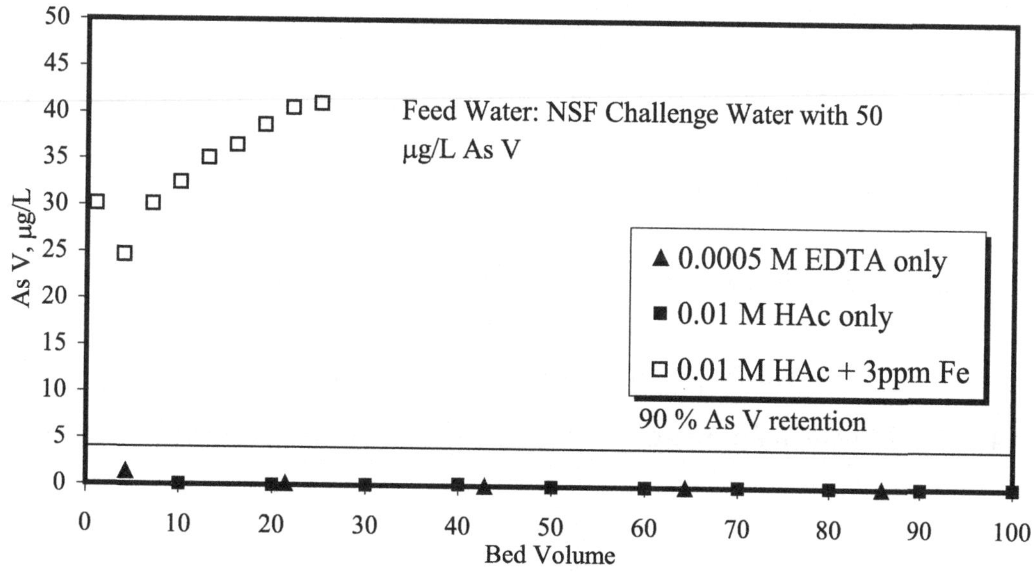

Figure 3.46 Breakthrough for 50 µg/L As V through IX-Cl using 7-mL of 50-100 Dowex resin with EDTA and HAc used separately with and without Fe II

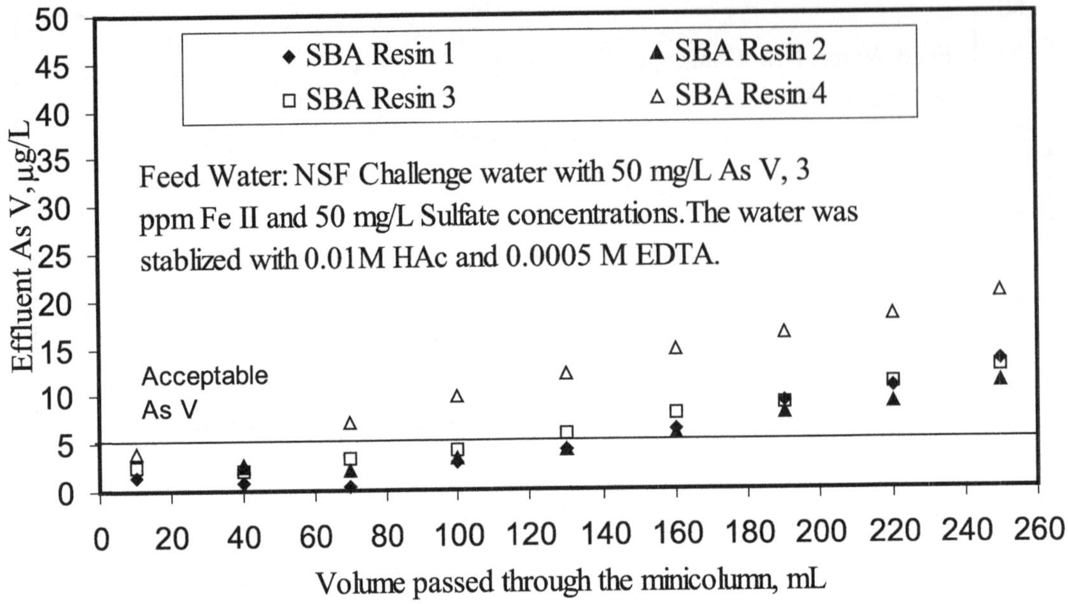

Figure 3.47 Breakthrough for 50 µg/L As V through 10 mL of four different resin types

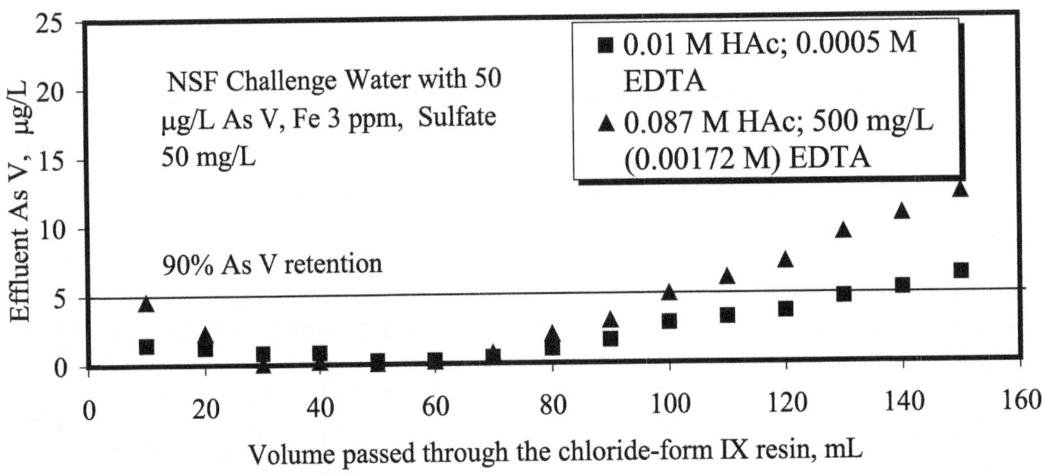

Figure 3.48 Breakthrough for 50 µg/L As V through 10-mL IX-Cl 50-100 Dowex resin

Field Speciation and Preservation Tests

Objectives:
(1) Validate field speciation method using HAc-EDTA and Cl-form ion-exchange method.
(2) Compare immediate field-speciated As III with next-day lab speciated As III on HAc-EDTA preserved sample.
Locations: 6 wells near Columbus, OH; 4 wells in Albuquerque, 2 wells in McFarland, CA; 2 wells in Grand Forks, ND. 12 wells in Minnesota. Replicates (3, typically) of samples are analyzed.

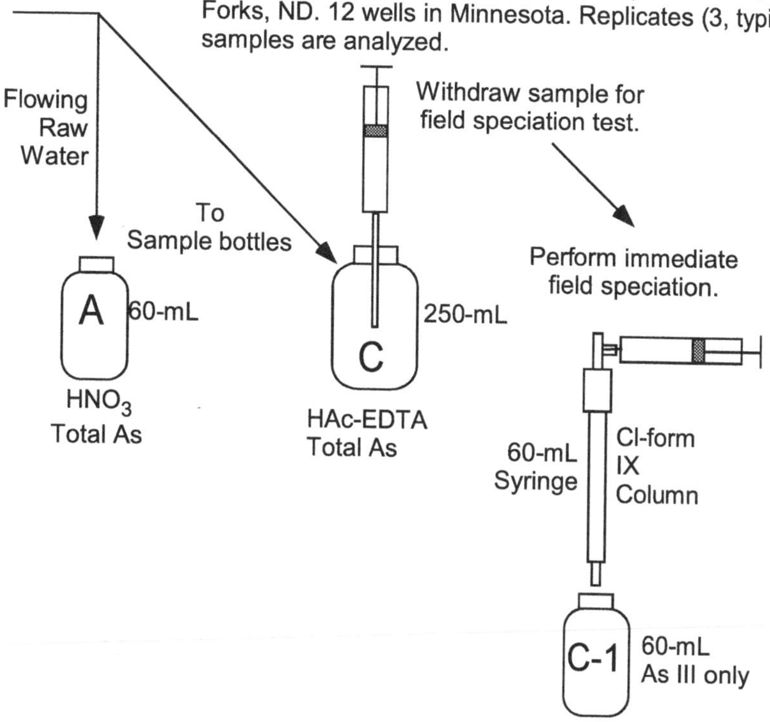

A: 60-mL total As sample preserved in HNO_3 @ pH 2. Shipped to UMinn for analysis of total arsenic by ICP-MS. Bottle A was not used in other locations.
C: 250-mL HAc-EDTA stabilized sample for field speciation. Preserved sample remaining after field speciation is sent to UH for total As analysis by HGAA and lab speciation of As III by pH 5.5-hydride generation.
C-1: Field speciated sample. HAc-EDTA stabilized, field-speciated As III sample shipped to UH by overnight mail for analysis by HGAA.
Total As Comparison: Total As measured by ICP-MS on Bottle A @ UMinn compared with total As measured on Bottle C by HGAA @ UH.
AsIII Comparison: Field speciatd As III measured by HGAA on Bottle C-1 at UH is compared with As III speciated in the UH lab from Bottle C using pH 5.5-hydride generation.

Figure 3.49 Schematic of field speciation kit and sampling procedure

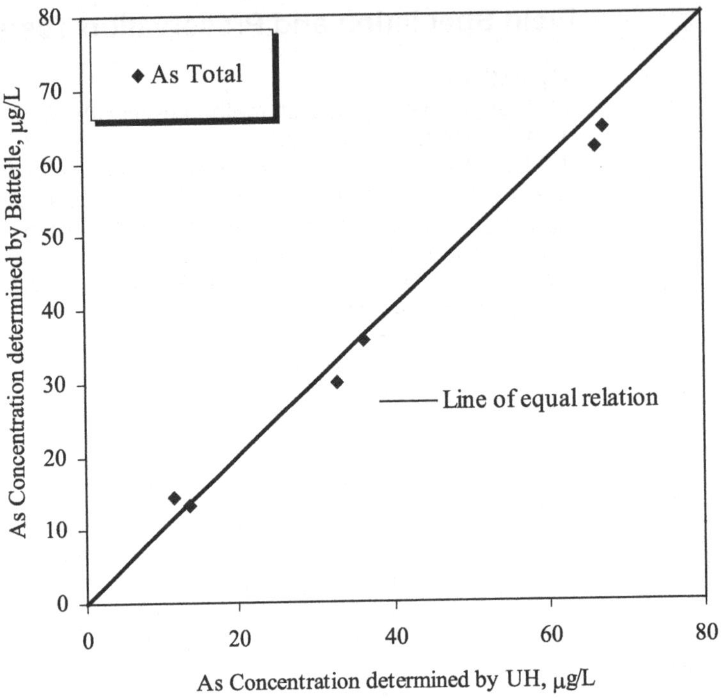

Figure 3.50 Comparison of UH and Battelle As Total analyses of field samples from Columbus area

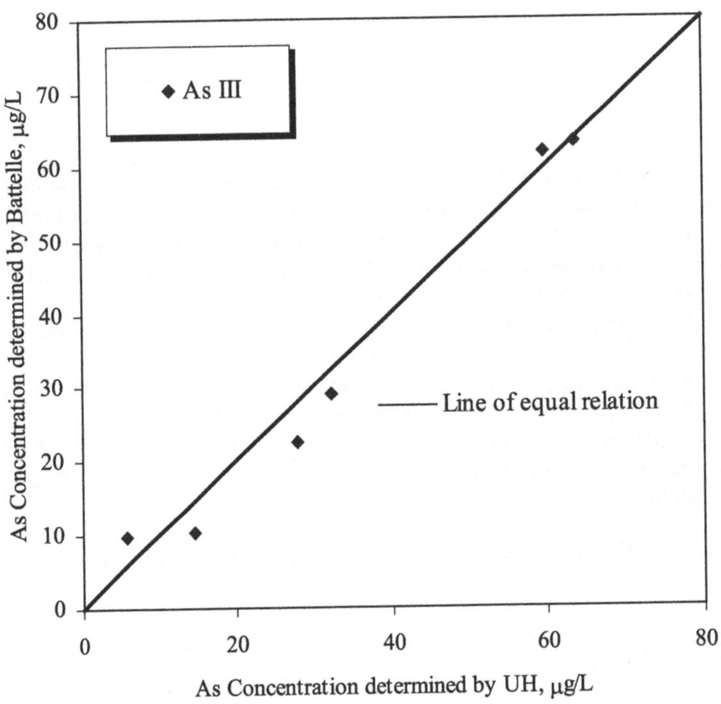

Figure 3.51 Comparison of UH and Battelle As III analyses of field samples from Columbus area

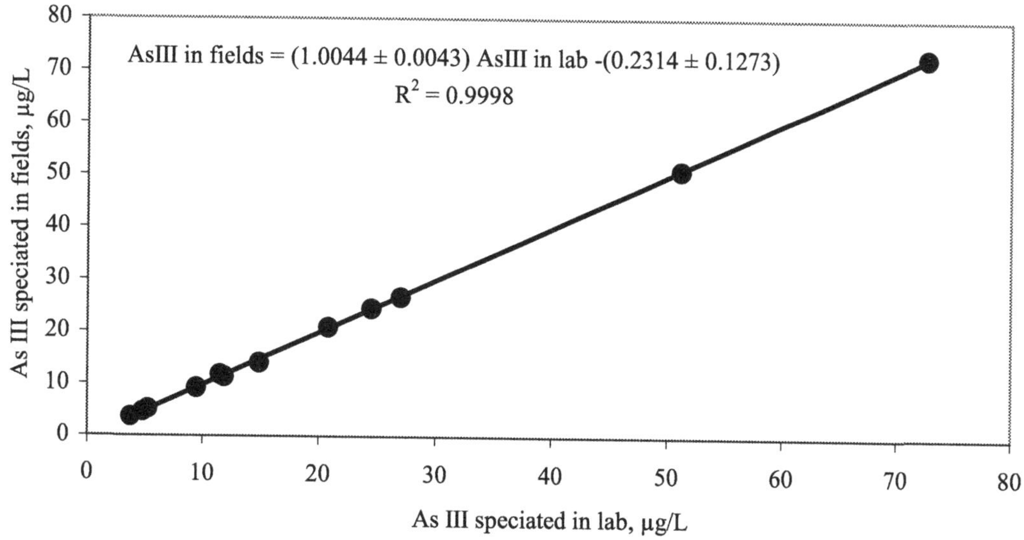

Figure 3.52 Correlation between the concentrations of As III in the preserved samples speciated in the field and As III measured in the laboratory by pH depended HG-AAS

CHAPTER 4
SUMMARY AND CONCLUSIONS

Laboratory-scale studies were performed at the University of Houston to develop a field speciation method for As III and As V. The study was divided into four phases: (1) preliminary experiments to evaluate the advantages and disadvantages of filter disks vs resin minicolumns and choice of resin form (chloride vs acetate) in the absence of ferrous iron, (2) evaluation of three previously-reported speciation methods in comparison with a new method based on sample pretreatment with acetic acid and EDTA to overcome iron interference, (3) rigorous testing of the chosen EDTA-HAc method to evaluate its accuracy and robustness, and (4) limited field verification tests of the new EDTA-HAc method at 23 different wells in five states: Ohio (6), North Dakota (2), New Mexico (2), California (1), and Minnesota (12).

The conclusions of the speciation of As III and As V study are summarized below.

1. IX filter disks lacked the capacity for providing at least 60 mL of throughput for the acidified samples (pH 2-3) where acidification was necessary to prevent iron interference.
2. In the absence of ferrous iron, acetate- and chloride-form resin minicolumns were equally efficient for arsenic speciation for unacidified samples. However, for acidified samples (pH 2.0 and 3.0), the acetate-form resin buffered the speciation pH in the 3.5-5 range and was able to retain As V, whereas the chloride-form resin provided no buffering and failed to effectively separate As III and As V.
3. The three existing methods evaluated—the pH 8.3- and 3.0- Clifford methods and the Edwards Method—were unable to overcome interferences due to ferrous iron oxidation and precipitation during sampling and speciation. The Edwards method, which includes a 0.45 mm filtration step to separate particulate arsenic, exposed the challenge water to air for a period of up to 10 min, which resulted in almost complete oxidation and precipitation of ferrous iron. If the filtration step is eliminated from the Edwards method, and samples are quickly acidified to pH <2, it is likely that the modified method will overcome ferrous iron oxidation and precipitation.
4. The new EDTA-HAc method was capable of speciating inorganic arsenic in the presence or absence of 3 mg/L ferrous iron because it immediately lowers the sample pH using HAc and complexes the iron using EDTA. As III and As V were efficiently separated on the chloride-form resin, which was not affected by flow rate or background ions in the water.
5. The new EDTA-HAc method, which utilizes a syringe to force sample through the resin minicolumn is faster than gravity-flow methods, and the new method does not use hazardous chemicals such as H_2SO_4 or HCl for acidification of samples prior to speciation.
6. The determination of particulate arsenic by filtration in the atmosphere was not feasible when ferrous iron was present in the nitrogen-sparged challenge water due to inevitable oxidation and precipitation of $Fe(OH)_3(s)$, which adsorbed both As V and As III.
7. In the presence of Fe II, HAc and EDTA, the run length of the chloride-form minicolumn is less than 10 bed volumes. However, in spite of short run length, the use of EDTA and HAc is required to avoid ferrous iron oxidation and subsequent precipitation.

8. At least 10 mL of chloride-form SBA resin must be used to obtain 60 mL of speciated sample and 40 mL of excess capacity as a safety factor for the EDTA-HAc method.
9. The EDTA-HAc and Edwards-Battelle methods produced similar speciation results on analysis of six field samples from the Columbus Ohio area.
10. For the twelve Minnesota well water samples, the concentrations of field speciated As III and concentrations of As III in the EDTA-HAc preserved samples speciated in the laboratory showed excellent correlation and no significant difference at the 95% confidence level.
11. The overall objective of the arsenic speciation study was attained, i.e., a robust field arsenic speciation method was developed to separate As III from As V for subsequent analysis of speciated As III and total As in an analytical laboratory.

APPENDIX A
SPECIATION MEDIA PREPARATION

IX FILTER DISK PRECONDITIONING

1. Center the filter on the base of a vacuum filter holder.
2. Wet the filter with 10 mL of acetone for 30 sec, and then apply vacuum to dry the filter.
3. Add 10 mL of isopropanol and allow filter to soak for 30 sec. Apply vacuum to dry the filter.
4. Finally, add 10 mL of methanol. Draw approximately 3-4 mL of methanol through the filter under vacuum. Vent vacuum and allow filter allowed to soak in methanol for 60 seconds. Reapply vacuum and add 30 mL of reagent grade water to rinse methanol from the filter. Ensure that the filter does not dry by leaving 3-5 mL of reagent grade water on the filter.

Note: This preconditioning procedure was found to be unnecessary as a preconditioning step for arsenic speciation using ion-exchange filter disks.

IX FILTER REGENERATION—UH METHOD

Preconditioned and used filters were regenerated according to the procedure outlined below.

1. Mount the filter on the base of a vacuum filtration apparatus and place a 0.22 μm filter (Millipore, 47 mm, Type GS) on top of the IX Empore filter. Add 20 mL of 1 M HCl and pull approximately 3-4 mL of HCl through the filter under vacuum. Vent the vacuum and allow the filter to soak for about 60 seconds. Reapply vacuum to draw the remaining acid through the filter at a flow rate of 20 mL/min or less.
2. (6) Add 20 mL of 1 M NaCl and pull approximately 3-4 mL of the salt solution through the filter. Vent the vacuum and allow the filter to soak for about 60 seconds. Reapply vacuum and draw the remaining salt solution through the filter at a flow rate of 20 mL/min or less.
3. (7) To rinse the concentrated NaCl out of the filter, add 20 mL of 0.005 M NaCl and draw about 15 mL through the filter at a flow rate of 20 mL/min or less. The IX filter is now ready to use for the As(III)/(V) speciation tests.
4. (8) Once preconditioned and regenerated, the filter should not be allowed to dry for an extended period of time. One way to prevent drying is to place the filter in a beaker or a sealed plastic bag containing 0.005 M NaCl solution.

ACETATE-FORM RESIN PREPARATION (EDWARDS METHOD)

1. Place 1 kg of 50-100-mesh resin[1] was into a 2-L beaker.
2. Add 1 L of 1 N NaOH to the resin and stir for 1 hour. Repeat this sodium hydroxide rinse 3 times in sequence.

---

1. Dowex 1X8 50-100-mesh resin, Supelco, Bellefonte, PA.

3. After rinsing with two 1-L batches of reagent grade water, rinse the resin in acetic acid. Add acid (1 L of 1 N reagent-grade acetic acid), and stir for 5 minutes, and drain. Repeat the acetic acid rinse three times in sequence.
4. Finally, rinse the mixture with three sequential 1-L batches of reagent grade water.

APPENDIX B
PREPARATION OF DO-FREE NSF CHALLENGE WATER

A 4-L carboy was filled with arsenic (As III or V) spiked-NSF-challenge water, taking care to ensure that there were no air bubbles trapped before closing it with a screw-on threaded cap. The cap had 3 bulkhead fittings; two to allow the entry and exit of the nitrogen gas and one to allow the spiking of iron to test waters with iron. At predetermined intervals, water from the carboy was allowed to exit and the DO concentration was determined using Hach method 8316 (Indigo Carmine Method) which has a method detection limit of 0.01 mg/L (10 μg/L). Sparging was continued until the DO measurements were below 0.05 mg/L residual dissolved oxygen (DO) (as recommended by NSF protocols for preparing NSF Challenge Water containing reduced iron or manganese). Such low concentrations were achieved after an affective sparging of 60 minutes (one hour). Schematic of the apparatus is shown in Figure B.1.

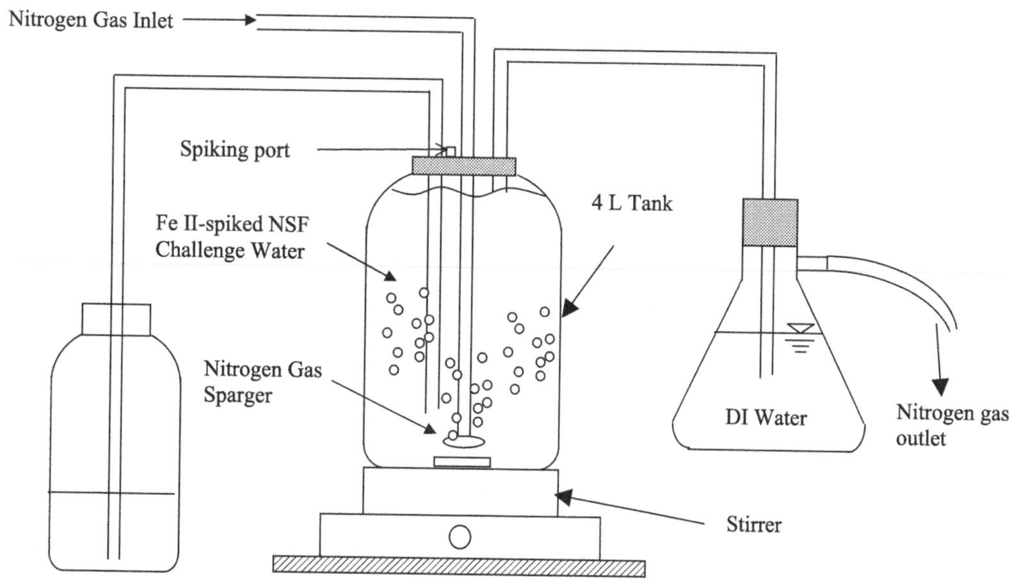

Figure B.1 Schematic diagram for preparation of DO-free challenge water

APPENDIX C
LOW-pH ACETATE-FORM IX FILTER DISKS

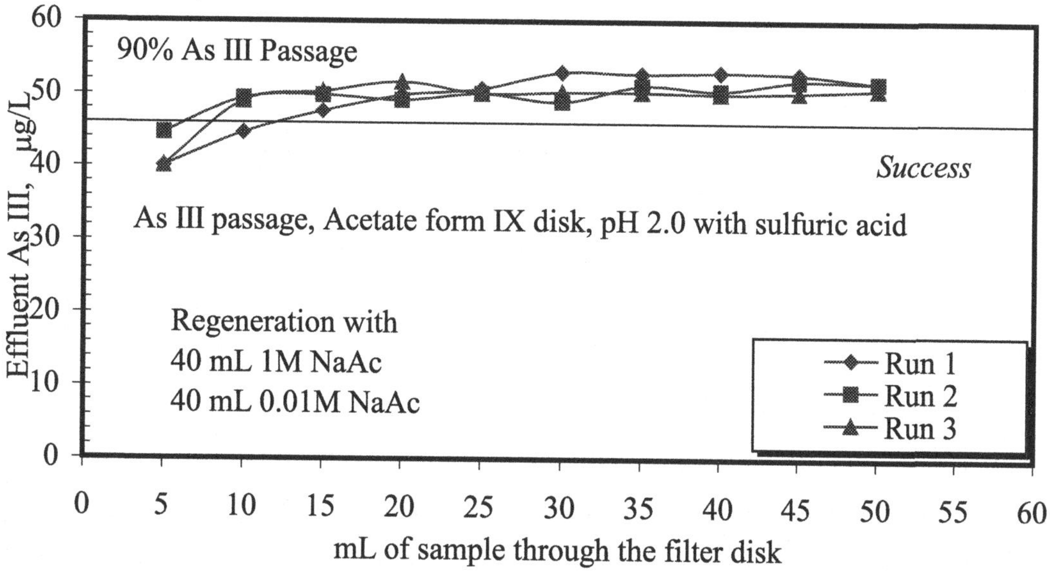

Figure C.1 Arsenic III passage through acetate-form IX filter disk with subsequent regeneration and reuse, feed pH = 3.0 using sulfuric acid

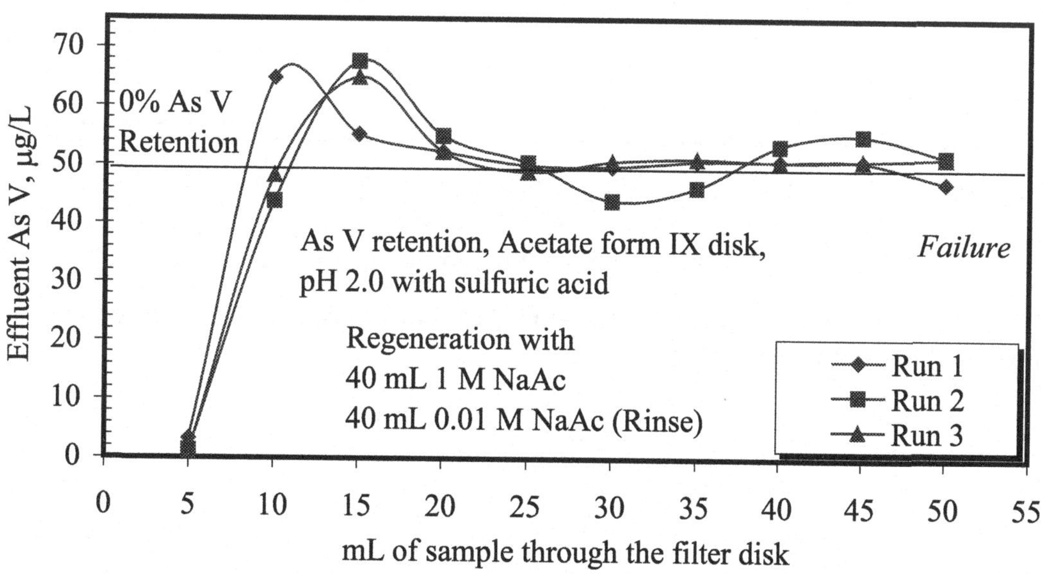

Figure C.2 Arsenic V retention through acetate-form IX filter disk with subsequent regeneration and reuse, feed pH = 2.0 using sulfuric acid

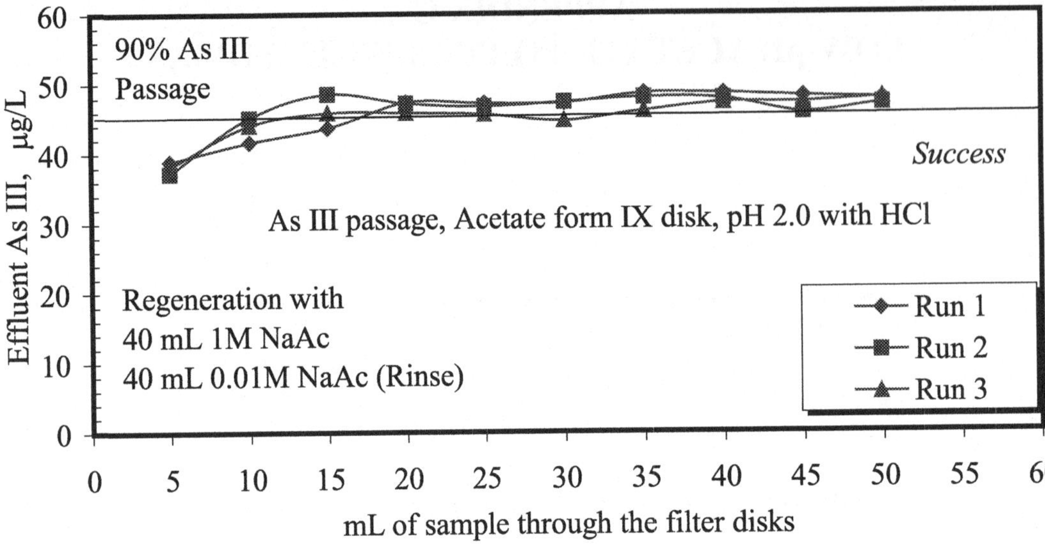

Figure C.3 As III passage through acetate-form IX filter disk with subsequent regeneration and reuse, feed pH = 2.0 using HCl

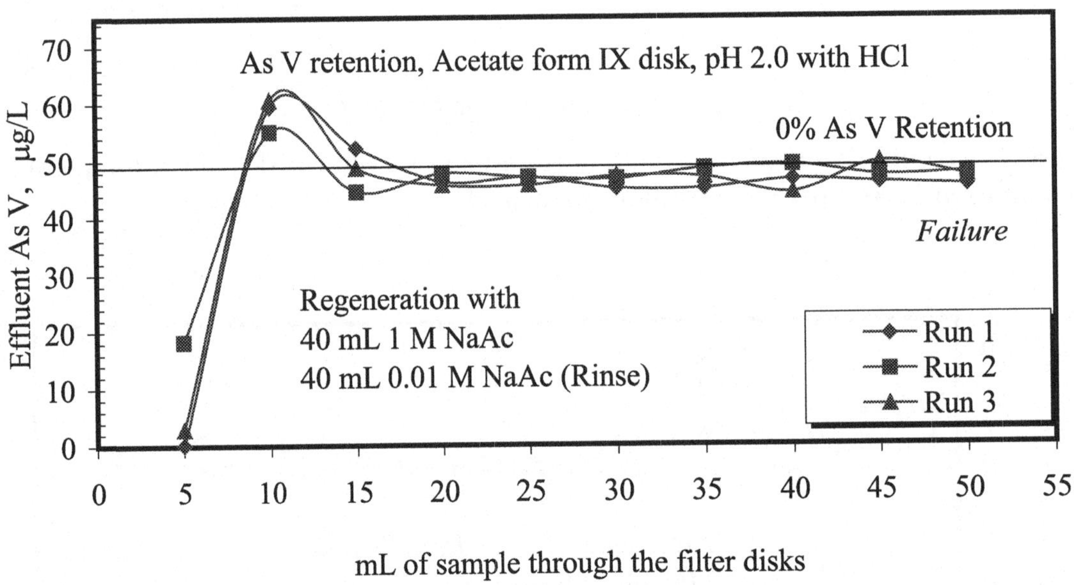

Figure C.4 As V retention through acetate-form IX filter disk with subsequent regeneration and reuse, feed pH = 2.0 using HCl

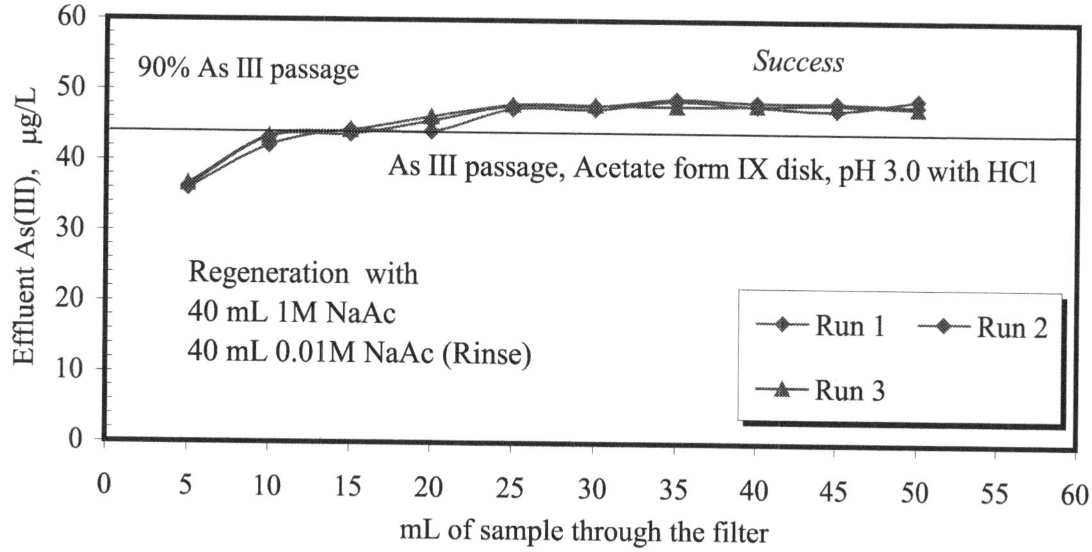

Figure C.5 As III passage through acetate-form IX filter disk with subsequent regeneration and reuse, feed pH = 3.0 using HCl

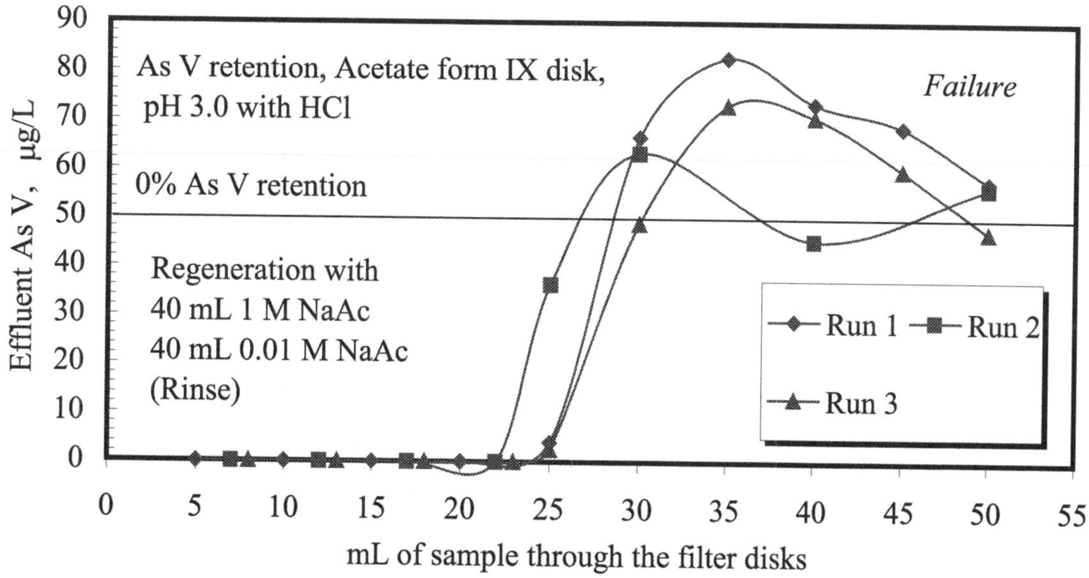

Figure C.6 As V retention on acetate-form IX filter disk with subsequent regeneration and reuse, feed pH = 3.0 using HCl

APPENDIX D
LOW-pH CHLORIDE-FORM RESIN MINICOLUMNS

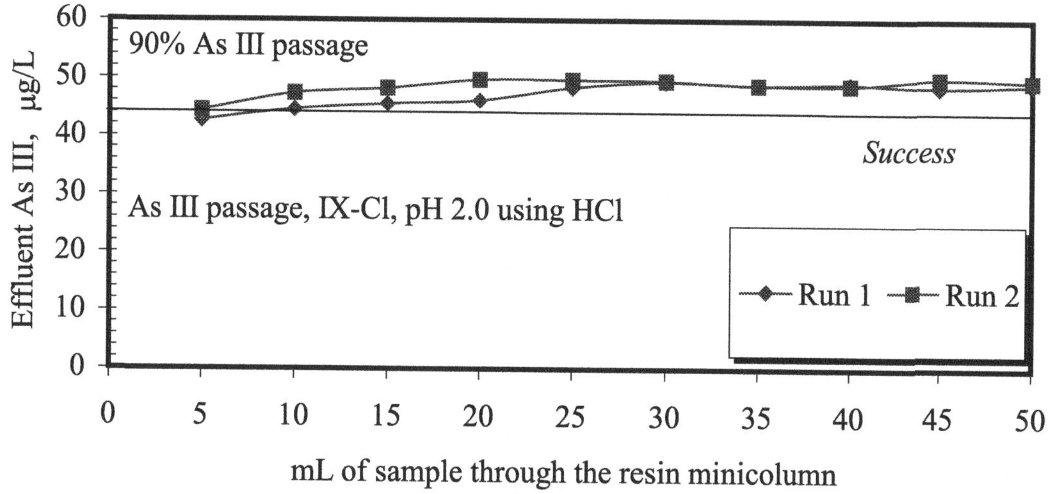

Figure D.1 Arsenic III passage through chloride-form IX minicolumn, feed pH = 2.0 using HCl

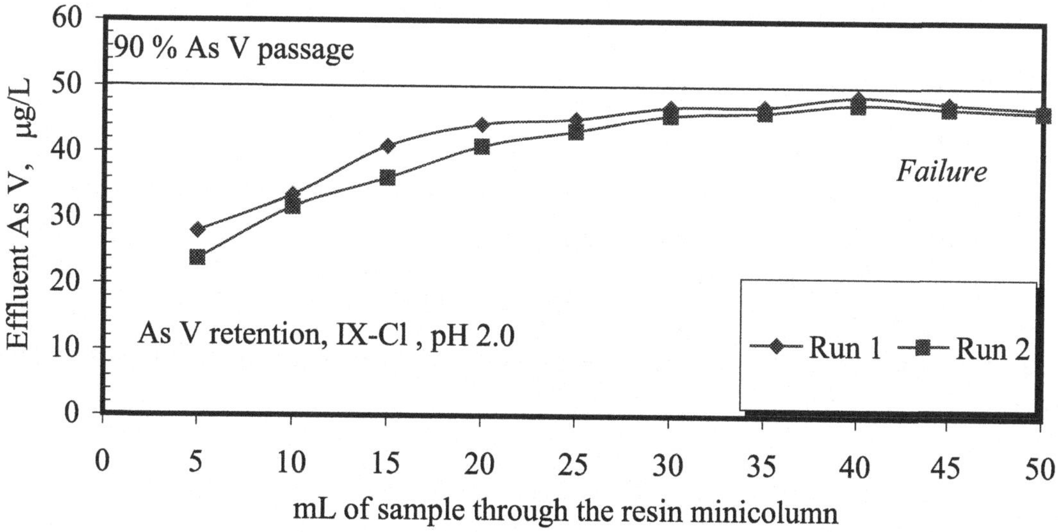

Figure D.2 Arsenic V retention on chloride-form IX minicolumn, feed pH = 2.0 using HCl

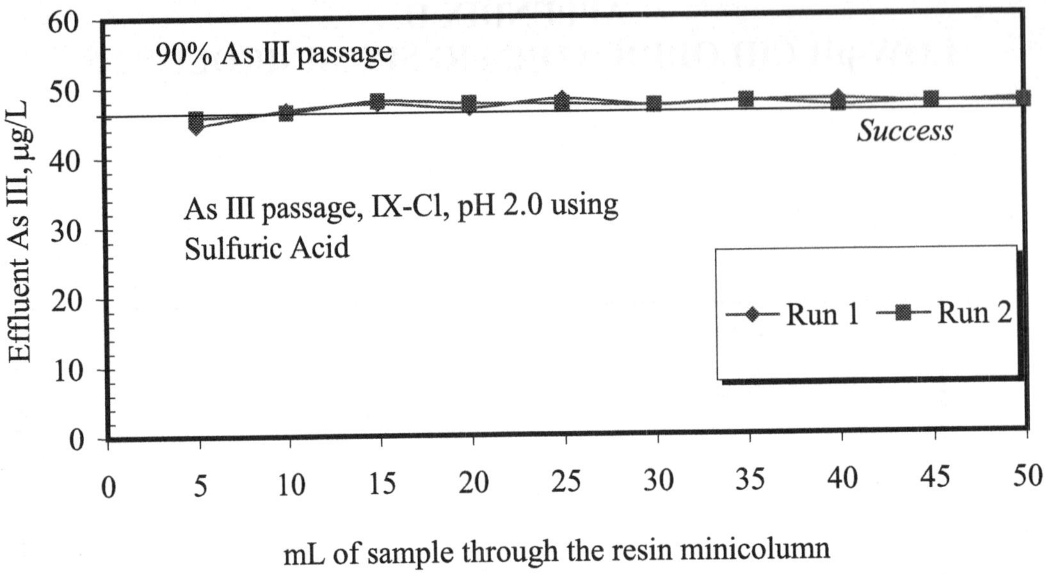

Figure D.3 Arsenic III passage through chloride-form IX minicolumn, feed pH = 2.0 using H_2SO_4

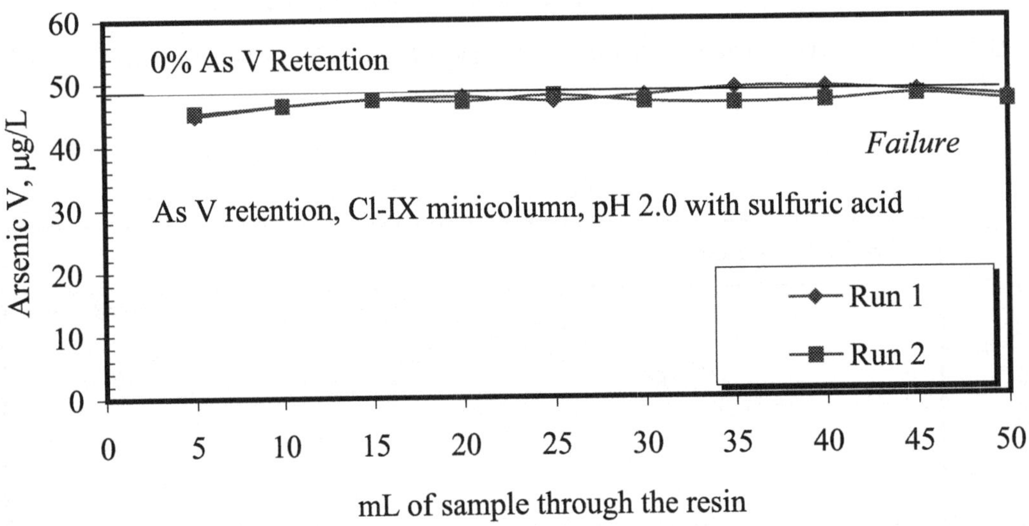

Figure D.4 Arsenic V retention on chloride-form IX minicolumn, feed pH = 2.0 using H_2SO_4

APPENDIX E
EFFECT OF EDTA AND HAc ON ARSENIC ANALYSIS USING HGAA

Table E.1
Composition of sample prepared for arsenic analysis to evaluate effect of added reagents (EDTA, HAc and iron)

	Volume of added reagent (mL)							
Spiked with	KI	HCl	As Std	HAc	EDTA	Fe II	DIW	Total
None	1.5	1.5	12.0	—	—	—	0.144	15.144
HAc	1.5	1.5	12.0	0.60	—	—	0.084	15.144
EDTA	1.5	1.5	12.0	—	0.60	—	0.084	15.144
Fe II	1.5	1.5	12.0	—	—	0.24	0.120	15.144
All the three	1.5	1.5	12.0	0.60	0.60	0.24	—	15.144

Table E.2
Absorbance of samples as analyzed using HGAA

Standards, μg/L	KI + HCl only	With added HAc	With added EDTA	With added Fe	HAc + EDTA + Fe
0	0.004	0.004	0.001	0.001	0.001
1.3	0.091	0.088	0.090	0.089	0.087
2.8	0.190	0.189	0.189	0.190	0.187
6.8	0.443	0.432	0.447	0.446	0.437
11	0.690	0.694	0.700	0.691	0.690

Calibration equations were obtained from Tables E.1 and E.2 and the regression analysis was performed. Table E.3 compares the equations and regression values for different matrices analyzed.

Table E.3
Regression results for the study of matrix effect on HGAA arsenic analysis

Matrix composition	Calibration equation	R^2
With KI + HCl only	Y = 0.0624x + 0.0104	0.9994
With added HAc	Y = 0.0625x + 0.0076	0.9998
With added EDTA	Y = 0.0635x + 0.0072	0.9995
With added Fe	Y = 0.0627x + 0.0087	0.9992
HAc + EDTA + Fe	Y = 0.0625x + 0.0066	0.9997

From Table E.3, it can be seen that the slope of the calibration equation doesn't depend on the composition of the matrix.

APPENDIX F
ANALYSIS OF DMA AND MMA USING HGAA FIAS UNIT

Table F.1
Fate of organoarsenicals during inorganic arsenic speciation, samples analyzed using HGAA

	Spiking				Analyzed As	
As species	As III	As V	DMA	MMA	Feed	Speciated effluent
As V + DMA	—	50	20	—	53.5	2.6
	—	50	20	—	52.2	2.8
As V + MMA	—	50	—	20	75.6	19.8
	—	50	—	20	76.3	20.1
As III + DMA	50	—	20	—	52.2	49.9
	50	—	20	—	53.0	51.2
As III + MMA	50	—	—	20	72.1	71.0
	50	—	—	20	72.3	72.6

Note: All concentrations are expressed in µg/L.

From Table F.1, it can be observed that DMA was not analyzed in the effluent coming out of the resin minicolumn. MMA passes through the resin minicolumn and will be measured and reported for As III.

APPENDIX G
ARSENIC FIELD SPECIATION KIT AND FIELD SPECIATION PROCEDURE

CONTENTS OF ARSENIC SPECIATION SAMPLING KITS

Each arsenic field speciation sampling kit contains:

1. One Ion-Exchange minicolumn with 10 mL chloride-form resin.
2. One 250-mL bottle marked "C" containing EDTA-HAc solutions.
3. One 60-mL bottle marked "C-1," speciation column effluent.
4. One 60-mL syringe with Luer-lock tip.
5. One 6-inch long piece of 1/8-inch flexible tubing.
6. One 3-way Luer Lock valve.

PREPARATION OF RESIN MINICOLUMNS

a. Rinse Kontes Flex Glass[1] minicolumn with DI water by removing the screw-on cap and the end plugs.
b. Make a slurry of about 15-mL resin[2] in 15-mL DI water.
c. Unscrew the cap of the minicolumn and remove the lower end plug. Pour the resin slurry in the minicolumn. Allow water to drain and let the resin level rise to about 1 cm over the thin neck of the column (refer to the Figure G.1).
d. Replace the lower-end plug, the screw-cap and the upper end plug. The resin minicolumn is ready to be used for speciation.

PREPARATION OF REAGENTS AND SAMPLE BOTTLES

a. Prepare a 2.0 M HAc by adding 11.5 mL of 17.4 M Glacial Acetic Acid[3] to 50 mL DI water and dilute to 100 ml with DI water.
b. Prepare a 0.1 M EDTA from disodium salt of EDTA[4] (FW 372.24) by adding 3.72 gm of salt into a small volume DI water. Mix to dissolve and make up to 100 mL with DI water).
c. Add 10.8 mL of 2.0 M HAc and 4.25 mL of 0.1 M EDTA to 250-mL bottle and label it Bottle C. This will be 0.087 M HAc and 0.0017 M EDTA after the raw water sample has been used to fill the bottle.
d. Label an empty 50-mL bottle as C-1 to collect speciated As III sample.
e. Place the resin minicolumn, bottle C, and bottle C-1 in one-gallon zip-lock bag, along with a fresh 60-mL syringe to force sample through the minicolumn, and a 6-inch long piece of 1/8-inch flexible tubing to withdraw sample from 250-mL bottle. Also, place the sampling and speciation instruction handout in the bag.
f. Tightly seal the bag and label it "Arsenic Speciation Kit."

---

1. Kontes Economy Columns, ID 1 cm, Column Length 10 cm, Fisher Scientific
2. Dowex IX8 50-100 mesh size resin, Supelco, Bellefonte, PA
3. EM Science, Gibbstown, NJ
4. Aldrich Chemical Company, Milwaukee, WI

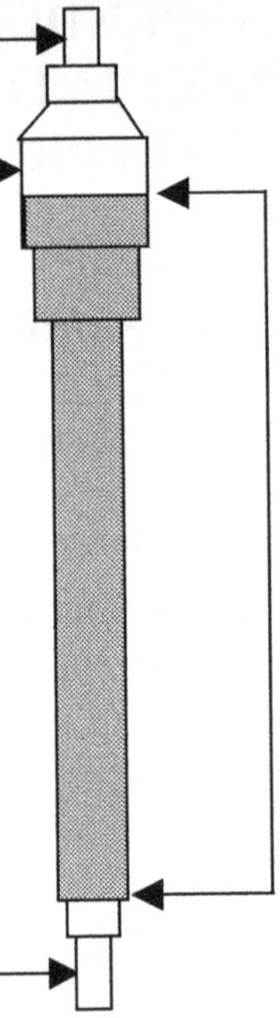

Figure G.1 Diagram of resin minicolumn (in the vertical position) showing its components and the level of resin

SAMPLING PROCEDURE FOR FIELD SPECIATION TESTS

Refer to Figures G.1, G.2, and G.3 when using the following procedure for field speciation of inorganic arsenic:

a. Flush all the sample lines thoroughly, and establish a flowing water sample through a flexible tube, typically 1/4 inch diameter.
b. Fill bottle C to a level just below its neck, being careful to avoid turbulence and air entrapment. Don't allow bottle C to overflow because it contains acetic acid and EDTA for pH adjustment and stabilization.
c. After capping bottle C, mix it thoroughly by inverting it 10 times.
d. Let bottle C sit for 2-3 minutes. During this time, unpack the 60-mL syringe, remove the cap on the syringe tip and slip the piece of 1/8-inch tubing provided over the syringe tip.
e. Grasp the resin minicolumn and remove both the end plugs. (The bigger end is the top of the column, i.e., the inlet where the syringe is to be connected.)
f. Insert the 1/8-inch tubing fitted to the syringe into the bottle C and withdraw 60 mL sample.
g. Remove the tubing from the syringe tip and twist the side port of the Luer-lock valve onto the syringe tip as shown in Figure G.3.
h. Mount the syringe (fitted with the Luer-lock valve and filled with sample from bottle C) onto the inlet of the resin minicolumn. See Figure G.3.
i. Force 60 mL of sample through the resin minicolumn during a period of 30 seconds to 3 minutes. Waste the first 15 mL of effluent from the minicolumn, and collect the last 45 mL effluent in the bottle labeled C-1.

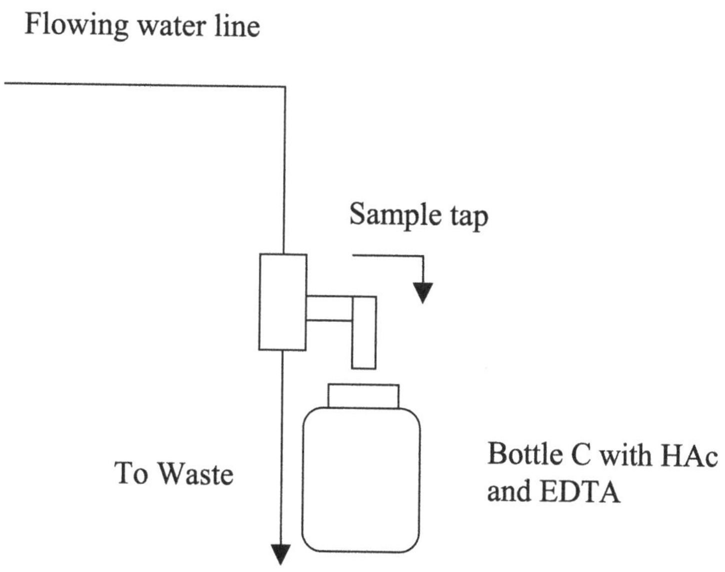

Figure G.2 Sample collection into bottle containing HAc and EDTA

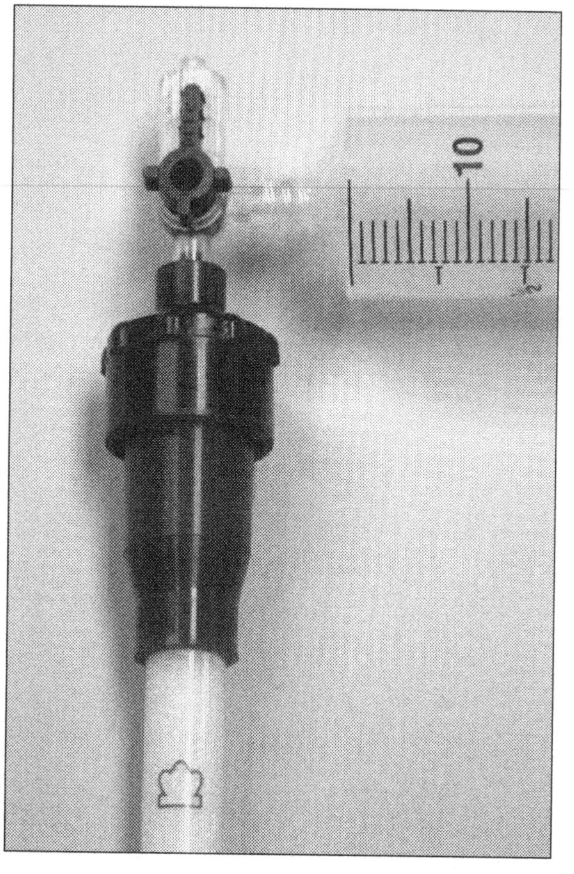

Figure G.3 Syringe mounted to 3-way Luer-lock valve with valve mounted to minicolumn inlet ("Off" position on valve is pointing upwards so that flow is from the syringe into the minicolumn. Flow leaving the minicolumn goes into bottle C-1.)

REFERENCES

Andreae, M. 1977. Determination of Arsenic Species in Natural Waters. *Anal Chem.*, 49(6):820–823.

Bately, G.E., and G.K.-C. Low. 1989. Application of High-Performance Liquid Chromatography to Trace Element Speciation Studies. In *Trace Element Speciation: Analytical Methods and Problems*. Boca Raton, Fla.: CRC Press.

Berthouex, P.M., and L.C. Brown. 1994. *Statistics for Environmental Engineers*. Lewis Publishers.

Clifford, D.A. 1999. Ion-Exchange and Inorganic Adsorption. In *Water Treatment and Disposal*. 5th ed. New York: McGraw-Hill.

Clifford, D.A., L. Ceber, and S. Chow. 1983. Separation of Arsenic III and Arsenic V by Ion-Exchange. In *Proc. of the 1983 AWWA Water Quality Technology Conference*. Denver, Colo.: AWWA.

Clifford, D.A., G. Ghurye, A. Tripp, and T. Jian. 1997. *City of Albuquerque Arsenic Study: Final Report, Phases 1 and 2*. Houston, Texas: University of Houston, Department of Civil and Environmental Engineering.

Clifford, D.A., and C.C. Lin. 1991. Arsenic III and Arsenic V Removal From Drinking Water in San Ysidro, N.M. EPA/600/S2/91/091. Cincinnati, Ohio: USEPA.

Clifford, D.A., and W.J. Weber Jr. 1983. The Determinants of Divalent/Monovalent Selectivity in Anion Exchangers. *React. Polym.*, 1(2):77–89.

Dean, J.A. 1992. *Lange's Handbook of Chemistry.* 14 ed. New York: McGraw-Hill.

Dudzinska, M., and D.A. Clifford. 1991. Anion Exchange Studies of Lead-EDTA Complexes. *React. Polym.*, 16(2):71–80.

Edwards, M., S. Patel, L. McNeill, H-W. Chen, M. Frey, A.D. Eaton, R.C. Antweiler, and H.E. Taylor. 1998. Consideration of As Analysis and Speciation. *Jour. AWWA*, 90(3):103–113.

Ferguson, J.F., and J. Gavis. 1972. A Review of the Arsenic Cycles in Natural Waters. *Water Res.*, 6(11):1259–1274.

Ficklin, W.H. 1983. Separation of As III and As V in Ground Waters by Ion-Exchange. *Talanta*, 30(5):371–373.

Gallagher, P.A., C.A. Schwegel, X. Wei, and J.T. Creed. 2001. Speciation and Preservation of Inorganic Arsenic in Drinking Water Sources Using EDTA With IC Separation and ICP-MS Detection. *J. Environ. Monit.*, 3(4):371–376.

Ghurye, G.L., and D.A. Clifford. 2001. Laboratory Study on the Oxidation of Arsenic III to Arsenic V. EPA/600/R-01/021.Washington, D.C.: USEPA.

Ghurye, G.L., D.A. Clifford, and A.R. Tripp. 1999. Combined Arsenic and Nitrate Removal by Ion Exchange, *Jour. AWWA*, 91(10):85–96.

Grabinski, A.A. 1981. Determination of As III, As V, MMA, DMA by Anion Exchange Chromatography With Flameless Atomic Absorption Spectrometric Detection. *Anal. Chem.*, 53(7):966–968.

Hering, J.G., and V.Q. Chiu. 2000. Arsenic Occurrence and Speciation in Municipal Ground-Water-Based Supply System. *J. Environ. Eng.*, 126(5):471–474.

Horng, L.L. 1983. Reaction Mechanisms and Chromatographic Behavior of Polyprotic Acid Anions in Multicomponent Ion-Exchange. Ph.D. diss., University of Houston, Texas.

Horng, L.L., and D.A. Clifford. 1997. The Behavior of Polyprotic Anions in Ion Exchange Resins. *React. Funct. Polym.*, 35l(1):41–54.

Hug, S.J., L. Canonica, M. Wegelin, D. Gechter, and U.V. Gunten. 2001. Solar Oxidation and Removal of Arsenic at Circumneutral pH in Iron Containing Water. *Environ. Sci. Technol.*, 35(10):2114–2121.

Irgolic, K.J. 1982. Speciation of Arsenic Compounds in Water Supplies. Project Summary. EPA-600/S1-82-010. Springfield, Va.: USEPA.

Johnson, W.F. 1958. *The Analytical Uses of Ethylenediaminetetraacetic Acid.* Princeton, N.J.: Van Nostrand.

Lopez, A., R. Torralba, M.A. Palacios, and C. Camara. 1992. Generation of AsH3 in the Absence of KI As Prereducing Agent: Speciation of Inorganic Arsenic. *Talanta*, 39(10):1343–1348.

NRC (National Research Council). 1977. *Drinking Water and Health.* Vol 1. Washington, D.C.: National Academy Press.

Reid, J. 1994. Arsenic Occurrence: USEPA Seeks Clearer Picture. *Jour. AWWA*, 86(9):44–51.

Subramonian, S., and D.A. Clifford. 1988. Monovalent/Divalent and the Charge Separation Concept. *React. Polym.*, 9(4):195–209.

Tong, J. 1997. Development of an Iron (III)-Coagulation-Microfiltration Process for Arsenic Removal in Groundwaters. Master's thesis. University of Houston, Texas.

ABBREVIATIONS

As	arsenic
As III	arsenic III
As V	arsenic V
AWWA	American Water Works Association
AwwaRF	Awwa Research Foundation
CA	California
DI	deionized
DMA	dimethylarsinic acid
DO	dissolved oxygen
EDTA	ethylenediaminetetraacetic acid
Fe II	iron II (ferrous iron)
Fe III	iron III (ferric iron)
FIAS	flow injection atomic spectroscopy
GFAAS	graphite furnace atomic absorption spectrophotometer
HAc	acetic acid (hydrogen acetate)
HGAA	hydride generation atomic absorption
ICP–MS	inductively coupled plasma–mass spectroscopy
IX	ion exchange
IXFD	ion-exchange filter disk
M	molar
MDL	minimum detection limit
meq	milliequivalent
meq/disk	milliequivalents per disk
mg/L	milligrams per liter
mL	milliliter
mL/min	milliliters per minute
MMA	monomethylarsonic acid
µg/L	micrograms per liter
ND	North Dakota
nm	nanometer
NM	New Mexico
NSF	National Sanitation Foundation

ppb	parts per billion
ppm	parts per million
ppmv	parts per million by volume
SAC	strong acid cation
SBA	strong base anion
USA	United States of America
USGS	United States Geological Survey

IWA Publishing's authorised EU representative for General Product Safety Regulations is Diane D'Arras, 15 rue Duret, 75116 Paris, France, e-mail: safety@iwap.co.uk.

Printed and bound by CPI Group (UK) Ltd, Croydon, CR0 4YY
17/09/2026
02218811-0009